JOSSEY-BASS TEACHER

Jossey-Bass Teacher provides educators with practical knowledge and tools to create a positive and lifelong impact on student learning. We offer classroom-tested and research-based teaching resources for a variety of grade levels and subject areas. Whether you are an aspiring, new, or veteran teacher, we want to help you make every teaching day your best.

From ready-to-use classroom activities to the latest teaching framework, our value-packed books provide insightful, practical, and comprehensive materials on the topics that matter most to K–12 teachers. We hope to become your trusted source for the best ideas from the most experienced and respected experts in the field.

The Algebra Teacher's Activity-a-Day

Grades 6–12

Over 180 Quick Challenges for Developing Math and Problem-Solving Skills

Frances McBroom Thompson

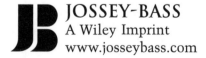

JOSSEY-BASS
A Wiley Imprint
www.josseybass.com

Published by Jossey-Bass
A Wiley Imprint
989 Market Street, San Francisco, CA 94103-1741—www.josseybass.com

Jossey-Bass books and products are available through most bookstores. To contact Jossey-Bass directly call our Customer Care Department within the U.S. at 800-956-7739, outside the U.S. at 317-572-3986, or fax 317-572-4002.

Jossey-Bass also publishes its books in a variety of electronic formats. Some content that appears in print may not be available in electronic books.

ISBN 978-0-4705-0517-5

Printed in the United States of America
FIRST EDITION
PB Printing 10 9 8 7 6 5 4 3 2 1

CONTENTS

ABOUT THIS BOOK

The Algebra Teacher's Activity-a-Day contains activities based on the content of Algebra I and II at the secondary level. Each activity may be used to supplement a daily algebra lesson by providing review of previous lessons or a focus for new lessons. Each activity emphasizes problem-solving strategies and logical reasoning, and often may have more than one solution; teachers should encourage students to communicate their different approaches or solutions both orally and in written form. The time required for most of the activities will be about five to ten minutes, depending on the type of activity selected and the amount of discussion encouraged. All activity pages are reproducible and may be copied for individual student use or projected on a screen for whole-class discussion.

The book is organized into ten sections containing fifteen to twenty activities per section, with a total of 180 activities. The sections are independent of each other and may be used in any order. Each section covers a wide range of topics. The activities within each section are ordered sequentially by algebraic content and by level of difficulty. The first page of each section gives general instructions as well as a sample activity with a possible solution. A grid that correlates each activity with the process and content standards developed by the National Council of Teachers of Mathematics (http://standards.nctm.org/document) appears before Section One. An answer key for all activities is provided at the end of the book.

Section One, "What Doesn't Belong?" offers experience with similarities and differences. Each activity presents four expressions or equations in a 2×2 grid. One expression or equation differs from the other three in some way. Each difference identified becomes a "solution" to the activity. Notation differences may be the focus of the activity, or procedural differences may be. Each activity has two or more possible solutions for students to discover.

Section Two, "What's Missing?" requires students to detect a change that has occurred between two expressions connected by an arrow. The arrow points to the result of the change. Another expression connected to a missing expression must also undergo the same change. Students must identify the missing expression to find the "solution." In some activities, the arrow may identify some element in the notation rather than a procedural change. A pair of arrows in an activity may represent a variety of relationships, thereby creating multiple solutions.

Section Three, "Where Is It?" provides activities in which students must locate a specific box in a grid of nine boxes. The item in the selected box must satisfy all of the clues given in the activity. The item might be an algebraic expression or equation, or a curve or set of curves. The process of elimination must be applied and the clues assist students in clarifying various mathematical definitions.

Section Four, "Algebraic Pathways," includes activities in which algebraic expressions must be simplified or equations or inequalities must be solved. To find an answer, students must draw a path through several boxes in a grid, beginning at the top of the grid. Each box contains a possible step that may or may not belong in the chosen simplification or solution process. The purpose is to draw a path that leads directly to an answer to be recorded below the grid, and the path must avoid unnecessary reversal of any steps. These activities encourage students to be more efficient in mathematical procedures. Several approaches are possible for solving the same problem, thereby producing several different pathways and increasing students' flexibility of thought. Each pathway found is considered a "solution" to the activity.

Section Five, "Squiggles," contains activities that consist of networks of connected points. Students must assign terms (algebraic expressions or equations) from a set to points in a network, or squiggle, so that any two connected terms satisfy a given rule or relationship. Each term must be uniquely assigned to a point; a successfully completed assignment of terms forms a "solution" for a squiggle. Different solutions are possible by varying the terms assigned to the points. This type of activity provides practice in analysis and logical reasoning, as well as review of definitions, factoring, and characteristics of graphs.

Section Six, "Math Mystery Messages," involves math definitions and properties. Students need much review of these theoretical topics. Although the activities appear to involve simple decoding, only a few number-letter pairs are provided as clues in each activity. To discover each message, students must apply logical reasoning, trial-and-error strategies, and understanding of the structure of the English language. Students for whom English is a second language, as well as students weak in math vocabulary, will find these activities difficult, but they will profit from the challenge.

Section Seven, "What Am I?" offers activities that contain sets of clues. Students must apply all of the clues in an activity to identify the expression that is the "solution." Deductive reasoning and review of math vocabulary are emphasized in this section.

Section Eight, "Al-ge-grams," requires students to apply accurately the order of operations and other mathematical procedures in order to simplify algebraic expressions. Once an activity's expression is simplified, the remaining letters, and perhaps numbers, must be unscrambled to form a special message. The message will be general, not necessarily mathematical.

Section Nine, "Potpourri," contains three types of activities: *cooperative games*, which allow students to work with a partner to solve nonroutine problems through hands-on activities; *oral team problems,* which involve teams of two to four students who must solve a problem only through oral discussion and mental mathematics—no calculators or paper and pencil allowed; and *mini-investigations,* which may be worked on by individuals or in small groups of students. Emphasis is on the use of counterexamples, number patterns, and problem-solving strategies such as making tables or creating easier problems.

Section Ten, "Calculator Explorations," provides two types of activities: *applications,* which require students, either independently or with a partner, to use regular calculators to generate data in which to identify patterns or from which to draw conclusions; and *graphical explorations,* which have partners use graphing calculators to investigate changes in functions and in their graphs. Predictor equations may also be found to match a given set of data.

ABOUT THE AUTHOR

FRANCES M^{c}BROOM THOMPSON has taught mathematics at the junior and senior high school levels and has served as a K–12 mathematics specialist. She holds a bachelor of science degree in mathematics education from Abilene Christian University (Texas), a master's degree in mathematics from the University of Texas at Austin, and a doctoral degree in mathematics education from the University of Georgia at Athens. Frances has published numerous articles and conducts workshops for teachers at the elementary and secondary levels. She is author of *Hands-On Algebra! Ready-to-Use Games and Activities for Grades 7–12* (Jossey-Bass, 1998); *Math Essentials: Middle School Level* (Jossey-Bass, 2005) and *High School Level* (Jossey-Bass, 2005); and *Five-Minute Challenges for Secondary School,* Volumes I and II (Activity Resources, 1988 and 1992).

ACKNOWLEDGMENTS

Special thanks are extended to the many classroom teachers, as well as to my graduate students in mathematics education, who have tested the ideas in this book over the last twenty years. Their suggestions for the activities have been extremely helpful and their enthusiasm has been encouraging.

This book would not have been possible without the continuous support of my husband, Claude, and our son, Brooks, who were so willing to share our one working computer during the preparation of the manuscript. Thanks, guys!

Appreciation is also extended to Jossey-Bass senior editor Kate Bradford and her senior editorial assistant Nana Twumasi for their assistance in the preparation of the final manuscript.

CORRELATION WITH NCTM PROCESS AND STANDARDS GRID

ACTIVITY BY SECTION

NCTM STANDARD		1. What Doesn't Belong?	2. What's Missing?	3. Where Is It?	4. Algebraic Pathways	5. Squiggles	6. Math Mystery Messages	7. What Am I?	8. Al-ge-grams	9. Pot-pourri	10. Cal-culator Explo-rations
PROCESS	Problem solving	1–20	1–20	1–15	1–20	1–20	1–20	1–15		1–7, 9–15	15
	Reasoning and proof	1–20	1–20	1–15	1–20	1–20	1–20	1–15	1–20	3–7, 9–15	1–15
	Communication	1–20	1–20	1–15	1–20	1–20	1–20	1–15	1–20	3–7, 9–15	1–15
	Connections			6, 13, 14, 15	1, 2	9	1–20	3, 12–15		4–8, 12,15	4–11
	Representation	1–20	1–20	1–15	1–20	1–20	1–20	1–15	1–20	4–7, 9–12, 15	1–15
CONTENT	Number properties	2, 3, 6, 9, 13	3, 4, 12	2, 3, 4, 5, 9, 10	13–17, 19	1, 2, 3, 14	1–8, 10–14	1, 2, 3, 4, 5, 6		1–14	1–15
	Add or subtract integers/reals	12, 14–18	3, 4, 11, 13, 15, 18, 19, 20	2, 7	1, 6–11, 14–20	1, 3, 6, 10, 11, 19, 20	2, 5, 13, 14, 19	3, 6	4–8, 12, 16–20	1–3, 6–9, 11–14	1–15
	Multiply or divide integers/reals	11–18	1, 2, 6–20	1, 2, 7	1, 2, 5–20	3, 4, 5, 6, 8, 10–20	3, 4, 6–12, 14, 17, 19, 20	4	1–20	8, 9, 11–14	1–15
	Apply absolute value		4	2, 3			5	3, 11		14	10
	Identify degree of polynomial expression	5, 6, 11, 12, 16–20		4, 5		7	18	4, 5, 9, 10			7–15
	Apply exponential properties	2, 3, 6, 8, 9, 10, 13, 14, 15, 18, 19	5–10, 13–20	4, 5, 9, 12	1, 3, 4, 5, 14–19	7, 8, 9, 12, 13, 16–20	6, 9, 10, 14	1, 7	1–8, 10–20	4–5, 13–15	2, 5, 11–15

CONTENT										
Add or subtract polynomials	11, 12, 14–19	5, 13, 15, 18, 20	9	6, 7, 8, 10, 11, 14–17, 19, 20	6, 10, 11	13, 17	6	4–20	4, 5, 9–11, 15	3, 5, 8–15
Multiply, factor, or divide monomials	1, 2, 4, 6, 8, 10, 12–19	1, 5, 6, 7, 10, 13–20	12	1–5, 10–20	5, 8–18, 20	3, 4, 8, 9, 11, 12, 17	2, 5, 6	1–5, 7–20	4, 5, 8, 9, 11, 14, 15	3, 5, 7–15
Multiply polynomials by distributive property	4, 12, 14, 15, 16, 19	2, 13, 14, 15, 18, 20	9	6, 7, 8, 9, 14–19		8, 17	8, 10	9–15, 17, 20	4, 5, 11, 12, 14, 15	3
Factor / divide polynomials	11, 12, 16, 17, 18, 19	10, 13, 16, 17, 19	10, 11	14–17, 19	14–18, 20	3	9	9–20	9, 11	3
Solve linear equations	5, 7	11, 12, 19	7, 8	2, 7, 8, 9, 12–19	10, 11		11		9–11, 13, 15	3, 6, 8, 9
Solve linear inequalities			8	10, 11						
Solve second-degree equations	5, 17, 20			1, 14, 15, 16, 17, 19	19, 20	4, 18	10		14, 15	7, 11, 12
Simplify rational expressions	4, 7, 8, 9, 13, 19	5, 9, 10	7	4, 5, 9, 12, 13				1–20	8	
Operate with rational expressions	4, 7, 8	5, 6, 10	7	4, 5, 9, 12, 13				1–20		
Solve proportions	4		7	12, 13						

ACTIVITY BY SECTION

NCTM STANDARD	1. What Doesn't Belong?	2. What's Missing?	3. Where Is It?	4. Algebraic Pathways	5. Squiggles	6. Math Mystery Messages	7. What Am I?	8. Algebrams	9. Potpourri	10. Calculator Explorations
Simplify numeric or algebraic radicals	3, 10	8, 14	12	15, 16	1, 4, 8, 12	6		10, 11, 12, 15, 17, 20		5
Solve equations involving radicals				15, 16, 18						5
Apply Pythagorean theorem	3		1						15	5
Solve systems of linear equations			13, 15	20					3	
Identify or evaluate functions or relations	5, 7, 20	18	6, 13, 14, 15	20	19, 20	15, 16, 19, 20	12, 13, 14, 15		10, 13	6, 8–15
Graph relations or functions	20		6, 13, 14, 15		10, 11, 19, 20	19, 20	13, 14, 15		10	6, 8–15
Identify change in functions (slope)			6, 13, 14, 15		10, 11	19, 20	12, 14		10, 13	6, 8, 9

CONTENT

SECTION ONE

What Doesn't Belong?

In the activities in this section, students must look for both similarities and differences among the four expressions or equations provided in a problem. Three will be alike in some way and the fourth will differ from the other three. Notation differences may be the focus of a problem, or similarities in mathematical procedures may be observed. Different relationships are possible, depending on which characteristic is noticed first. Students should be encouraged to find as many "solutions" (differences or similarities) as possible for each problem, always stating their reasons for each solution.

Example 1

Three of the following expressions—*a*, *b*, *c*, and *d*—have something in common. The other one differs from them in some way. Which one does not belong? Give a reason. Several answers may be possible, but for different reasons. Can you find more than one possible choice?

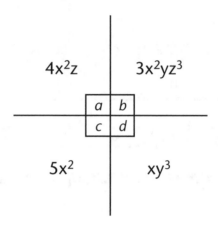

$$4x^2z \qquad\qquad 3x^2yz^3$$

$$\begin{array}{|c|c|} \hline a & b \\ \hline c & d \\ \hline \end{array}$$

$$5x^2 \qquad\qquad xy^3$$

Explanation: There are at least three possible solutions to this problem. Students may notice that expression *c* has only one variable whereas the other three expressions—*a*, *b*, and *d*—have at least two variables. So expression *c*, with "only one variable," would be considered a solution. Another possible solution would be expression *d*: it does not contain x to the second power, but the other three expressions do. Expression *d* also provides a third solution: its coefficient is $+1$ whereas the other expressions have coefficients not equal to $+1$. The emphasis in this particular problem is on the differences or similarities in the notations themselves and does not involve a mathematical procedure as such.

The answer key for this section provides several solutions for each problem. Other solutions may be possible, depending on the creativity of the students. These problems are effective in strengthening students' analytical skills. Reasons for choices should be shared during class discussion of the problems.

NAME _____   DATE _____

1.1

Three of the following expressions—a, b, c, and d—have something in common. The other one differs from them in some way. Which one does not belong? Give a reason. Several answers may be possible, but for different reasons. Can you find more than one possible choice?

ml   cm

a	b
c	d

kg   xm

_____

_____

_____

_____

_____

_____

_____

1.2

Three of the following expressions—a, b, c, and d—have something in common. The other one differs from them in some way. Which one does not belong? Give a reason. Several answers may be possible, but for different reasons. Can you find more than one possible choice?

$4x^2z$ $\qquad$ $3x^2yz^3$

a	b
c	d

$5x^2$ $\qquad$ xy^3

_____

_____

_____

_____

_____

_____

_____

_____

1.3

Three of the following expressions—a, b, c, and d—have something in common. The other one differs from them in some way. Which one does not belong? Give a reason. Several answers may be possible, but for different reasons. Can you find more than one possible choice?

3, 4, 5	6.9, $\sqrt{84.64}$, 11.5

a	b
c	d

8, 10, 12	5, 12, 13

_____

_____

_____

_____

_____

_____

_____

1.4

Three of the following equations—a, b, c, and d—have something in common. The other one differs from them in some way. Which one does not belong? Give a reason. Several answers may be possible, but for different reasons. Can you find more than one possible choice?

$$\frac{am}{bn} = \frac{ar}{bg}$$

$$\frac{ab}{ad} = \frac{cd}{cb}$$

a	b
c	d

$$\frac{ab}{ac} = \frac{be}{ce}$$

$$\frac{a+e}{c+e} = \frac{a-e}{c-e}$$

NAME _____   DATE _____

 1.5

Three of the following equations—*a*, *b*, *c*, and *d*—have something in common. The other one differs from them in some way. Which one does not belong? Give a reason. Several answers may be possible, but for different reasons. Can you find more than one possible choice?

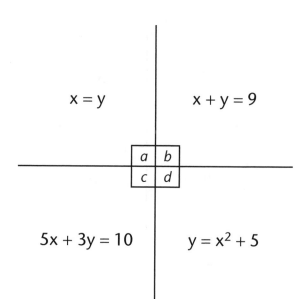

$x = y$ $x + y = 9$

| a | b |
| c | d |

$5x + 3y = 10$ $y = x^2 + 5$

_____

_____

_____

_____

_____

_____

1.6

Three of the following expressions—a, b, c, and d—have something in common. The other one differs from them in some way. Which one does not belong? Give a reason. Several answers may be possible, but for different reasons. Can you find more than one possible choice?

$3\sqrt{y}$ $4y^4$

a	b
c	d

$8x^3y$ $6x^2$

_____

_____

_____

_____

_____

_____

NAME _____ DATE _____

 1.7

Three of the following equations—*a*, *b*, *c*, and *d*—have something in common. The other one differs from them in some way. Which one does not belong? Give a reason. Several answers may be possible, but for different reasons. Can you find more than one possible choice?

Given: k, c constants
x, y variables

$$x = \frac{k}{y}$$

$$x = \frac{y}{c}$$

a	b
c	d

$$1 = \frac{c}{xy}$$

$$kx = \frac{c}{y}$$

_____

_____

_____

_____

_____

_____

1.8

Three of the following equations—*a*, *b*, *c*, and *d*—have something in common. The other one differs from them in some way. Which one does not belong? Give a reason. Several answers may be possible, but for different reasons. Can you find more than one possible choice?

$(x/y^4)^2 = x^2/y^8$	$(m^3/m^2)^2 = m^2$

a	b
c	d

$(m^8n^2)/p^6 =$ $[(m^4n)/p^3]^2$	$[(2a^4)/b^3]^3 =$ $(8a^{12})/b^9$

_____

_____

_____

_____

_____

_____

1.9

Three of the following expressions—*a*, *b*, *c*, and *d*—have something in common. The other one differs from them in some way. Which one does not belong? Give a reason. Several answers may be possible, but for different reasons. Can you find more than one possible choice?

$$\frac{x^{-6}}{x^7} \qquad \frac{1}{d^{13}}$$

a	b
c	d

$$\frac{y^{-3}}{y^{11}} \qquad \frac{c^{-9}}{c^4}$$

_____

_____

_____

_____

_____

_____

_____

1.10

Three of the following expressions—*a*, *b*, *c*, and *d*—have something in common. The other one differs from them in some way. Which one does not belong? Give a reason. Several answers may be possible, but for different reasons. Can you find more than one possible choice?

$$\sqrt[3]{a^5d^6c^2} \qquad a(\sqrt[3]{d^6a^2c^2})$$

a	b
c	d

$$ad^2(\sqrt[3]{a^2c^2}) \qquad d^2c(\sqrt[3]{a^5})$$

_____

_____

_____

_____

_____

_____

_____

NAME _____ DATE _____

1.11

Three of the following expressions—*a*, *b*, *c*, and *d*—have something in common. The other one differs from them in some way. Which one does not belong? Give a reason. Several answers may be possible, but for different reasons. Can you find more than one possible choice?

$3y + 2z$ $8z + 4x$

a	*b*
c	*d*

$3x^2 + 5x$ $18w^2 + 9x$

_____

_____

_____

_____

_____

_____

_____

Copyright © 2010 by John Wiley & Sons, Inc., *The Algebra Teacher's Activity-a-Day*

1.12

Three of the following expressions—*a*, *b*, *c*, and *d*—have something in common. The other one differs from them in some way. Which one does not belong? Give a reason. Several answers may be possible, but for different reasons. Can you find more than one possible choice?

$x^2 + 4x + 4$ $x^2 + 6x + 6$

a	*b*
c	*d*

$x^2 + 2x + 1$ $4x^2 + 12x + 9$

1.13

Three of the following expressions—*a*, *b*, *c*, and *d*—have something in common. The other one differs from them in some way. Which one does not belong? Give a reason. Several answers may be possible, but for different reasons. Can you find more than one possible choice?

$$4x^{-4}z \qquad\qquad 3ax^{-3}$$

a	b
c	d

$$\frac{2dx^{-2}}{c^{-2}} \qquad\qquad \frac{7c}{(yz)^{-4}}$$

1.14

Three of the following expressions—*a*, *b*, *c*, and *d*—have something in common. The other one differs from them in some way. Which one does not belong? Give a reason. Several answers may be possible, but for different reasons. Can you find more than one possible choice?

$(x + 3)^2$ $(x + y)^3$

a	b
c	d

y^{18} 27

Copyright © 2010 by John Wiley & Sons, Inc., *The Algebra Teacher's Activity-a-Day*

NAME _____   DATE _____

1.15

Three of the following expressions—*a*, *b*, *c*, and *d*—have something in common. The other one differs from them in some way. Which one does not belong? Give a reason. Several answers may be possible, but for different reasons. Can you find more than one possible choice?

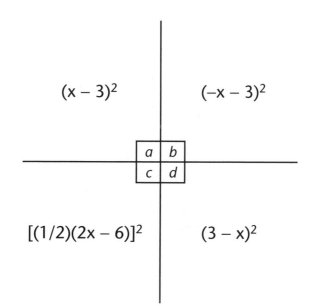

$(x - 3)^2$ $(-x - 3)^2$

a	b
c	d

$[(1/2)(2x - 6)]^2$ $(3 - x)^2$

_____

_____

_____

_____

_____

_____

1.16

Three of the following expressions—*a*, *b*, *c*, and *d*—have something in common. The other one differs from them in some way. Which one does not belong? Give a reason. Several answers may be possible, but for different reasons. Can you find more than one possible choice?

$x^2 + 6x + 9$ | $x^2 - 4x + 4$

| a | b |
| c | d |

$x^2 - 7x + 10$ | $4x^2 - 12x + 9$

NAME _____   DATE _____

 **1.17**

Three of the following equations—*a*, *b*, *c*, and *d*—have something in common. The other one differs from them in some way. Which one does not belong? Give a reason. Several answers may be possible, but for different reasons. Can you find more than one possible choice?

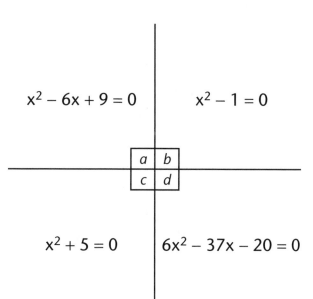

$x^2 - 6x + 9 = 0$ $x^2 - 1 = 0$

| a | b |
| c | d |

$x^2 + 5 = 0$ $6x^2 - 37x - 20 = 0$

_____

_____

_____

_____

_____

_____

1.18

Three of the following expressions—*a*, *b*, *c*, and *d*—have something in common. The other one differs from them in some way. Which one does not belong? Give a reason. Several answers may be possible, but for different reasons. Can you find more than one possible choice?

$$4x^2 + 36x + 81 \qquad 4y^2 + 4y + 1$$

a	*b*
c	*d*

$$4x^2 - 9x - 9 \qquad 4x^2 - 20x + 25$$

_____

_____

_____

_____

_____

_____

1.19

Three of the following expressions—*a*, *b*, *c*, and *d*—have something in common. The other one differs from them in some way. Which one does not belong? Give a reason. Several answers may be possible, but for different reasons. Can you find more than one possible choice?

$$\frac{a^4 - b^4}{a^2 + b^2}$$

$$\frac{a^2 + 2ab + b^2}{a + b}$$

a	b
c	d

$$\frac{a^3 - b^3}{a - b}$$

$$\frac{a^2 - ab - 2b^2}{a - 2b}$$

_____

_____

_____

_____

_____

1.20

Three of the following equations—*a*, *b*, *c*, and *d*—have something in common. The other one differs from them in some way. Which one does not belong? Give a reason. Several answers may be possible, but for different reasons. Can you find more than one possible choice?

$$\frac{x^2}{4} - \frac{y^2}{15} = 1 \qquad 16x^2 - 9y^2 + 144 = 0$$

a	b
c	d

$$\frac{x^2}{25} + \frac{y^2}{16} = 1 \qquad \frac{x^2}{4} - \frac{y^2}{9} = 1$$

_____

_____

_____

_____

_____

SECTION TWO

What's Missing?

In these activities, students must analyze two given parts of a problem connected by an arrow to determine what relationship exists between the two parts. They must then apply this same relationship to the third given part, which is connected to a fourth missing part by another arrow. The relationship may be an actual mathematical process, an equation paired with its solutions, or an expression matched to its factors or some power. The possibilities will vary greatly. Alternative answers may exist that are not listed in the answer key.

Example 2

In the following diagram, two algebraic expressions are being changed or related to new forms following the same procedure or process. The arrows point to the new forms. One space is empty. Can you decide what the procedure is and what should go in the empty space? State your reason. Other reasons may be possible. Can you find another?

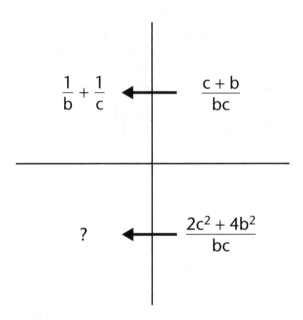

Explanation: Students should notice that the top two parts of the diagram involve the reversal of the addition process applied to two fractional forms. That is, the left-pointing arrow indicates that a sum has been transformed into two addends. So the bottom part with a left-pointing arrow must also represent a sum being transformed into two addends.

One possible way to find the two missing addends is to separate the bottom sum into its partial sums: $2c^2/bc$ and $4b^2/bc$. The two partial sums may then be simplified to produce the two addends: $2c/b$ and $4b/c$. The missing part has now been found: $2c/b + 4b/c$.

Students should be encouraged to share their analytical methods during a class discussion of the final solution. Alternative processes and solutions should be recognized as well.

2.1

In the following diagram, two algebraic expressions are being changed or related to new forms following the same procedure or process. The arrows point to the new forms. One space is empty. Can you decide what the procedure is and what should go in the empty space? State your reason. Other reasons may be possible. Can you find another?

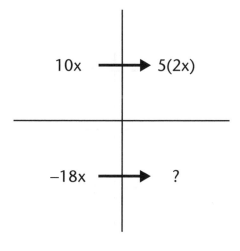

10x ⟶ 5(2x)

−18x ⟶ ?

_____

_____

_____

_____

_____

_____

_____

2.2

In the following diagram, two algebraic expressions are being changed or related to new forms following the same procedure or process. The arrows point to the new forms. One space is empty. Can you decide what the procedure is and what should go in the empty space? State your reason. Other reasons may be possible. Can you find another?

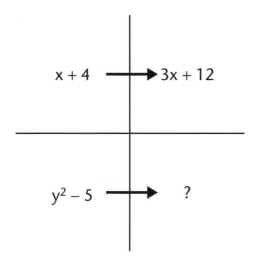

x + 4 ⟶ 3x + 12

$y^2 - 5$ ⟶ ?

_____

_____

_____

_____

_____

_____

Copyright © 2010 by John Wiley & Sons, Inc., *The Algebra Teacher's Activity-a-Day*

NAME _____   DATE _____

 **2.3**

In the following diagram, two algebraic expressions are being changed or related to new forms following the same procedure or process. The arrows point to the new forms. One space is empty. Can you decide what the procedure is and what should go in the empty space? State your reason. Other reasons may be possible. Can you find another?

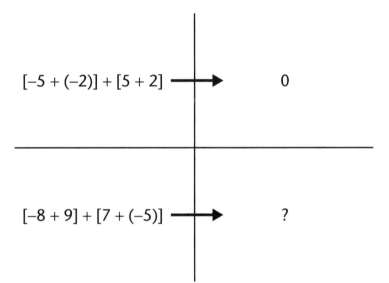

$[-5 + (-2)] + [5 + 2]$ ⟶ 0

$[-8 + 9] + [7 + (-5)]$ ⟶ $?$

_____

_____

_____

_____

_____

_____

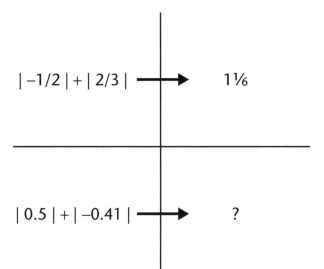

2.4

In the following diagram, two algebraic expressions are being changed or related to new forms following the same procedure or process. The arrows point to the new forms. One space is empty. Can you decide what the procedure is and what should go in the empty space? State your reason. Other reasons may be possible. Can you find another?

| −1/2 | + | 2/3 | ———▶ 1⅙

| 0.5 | + | −0.41 | ———▶ ?

NAME _____   DATE _____

 **2.5**

In the following diagram, two algebraic expressions are being changed or related to new forms following the same procedure or process. The arrows point to the new forms. One space is empty. Can you decide what the procedure is and what should go in the empty space? State your reason. Other reasons may be possible. Can you find another?

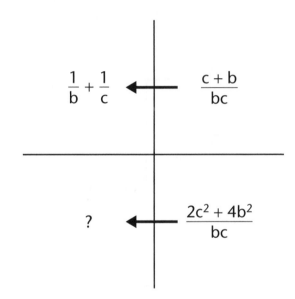

$$\frac{1}{b} + \frac{1}{c} \quad \longleftarrow \quad \frac{c + b}{bc}$$

$$? \quad \longleftarrow \quad \frac{2c^2 + 4b^2}{bc}$$

_____

_____

_____

_____

_____

_____

2.6

In the following diagram, two algebraic expressions are being changed or related to new forms following the same procedure or process. The arrows point to the new forms. One space is empty. Can you decide what the procedure is and what should go in the empty space? State your reason. Other reasons may be possible. Can you find another?

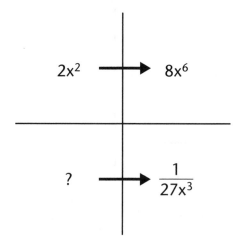

$$2x^2 \longrightarrow 8x^6$$

$$? \longrightarrow \frac{1}{27x^3}$$

NAME _____ DATE _____

 **2.7**

In the following diagram, two algebraic expressions are being changed or related to new forms following the same procedure or process. The arrows point to the new forms. One space is empty. Can you decide what the procedure is and what should go in the empty space? State your reason. Other reasons may be possible. Can you find another?

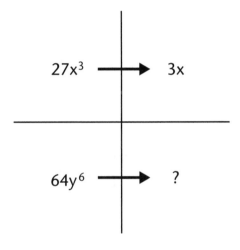

$27x^3 \longrightarrow 3x$

$64y^6 \longrightarrow$?

_____

_____

_____

_____

_____

_____

 **2.8**

In the following diagram, two algebraic expressions are being changed or related to new forms following the same procedure or process. The arrows point to the new forms. One space is empty. Can you decide what the procedure is and what should go in the empty space? State your reason. Other reasons may be possible. Can you find another?

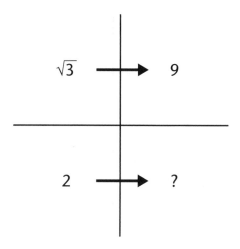

_____

_____

_____

_____

_____

_____

NAME _____   **DATE** _____

2.9

In the following diagram, two algebraic expressions are being changed or related to new forms following the same procedure or process. The arrows point to the new forms. One space is empty. Can you decide what the procedure is and what should go in the empty space? State your reason. Other reasons may be possible. Can you find another?

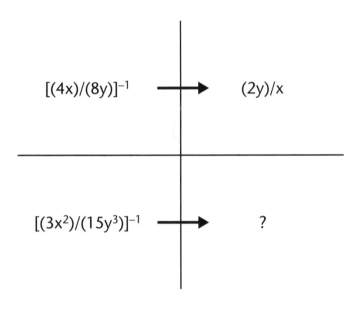

$[(4x)/(8y)]^{-1}$ ⟶ $(2y)/x$

$[(3x^2)/(15y^3)]^{-1}$ ⟶ ?

_____

_____

_____

_____

_____

_____

2.10

In the following diagram, two algebraic expressions are being changed or related to new forms following the same procedure or process. The arrows point to the new forms. One space is empty. Can you decide what the procedure is and what should go in the empty space? State your reason. Other reasons may be possible. Can you find another?

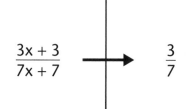

$$\frac{3x + 3}{7x + 7} \longrightarrow \frac{3}{7}$$

$$\frac{3x^2 + 6}{7x^3 + 14x} \longrightarrow \ ?$$

_____

_____

_____

_____

_____

_____

NAME _____   **DATE** _____

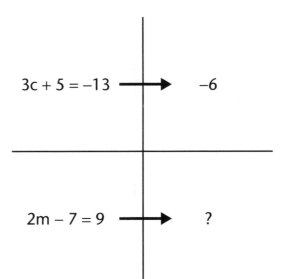

2.11

In the following diagram, two algebraic expressions are being changed or related to new forms following the same procedure or process. The arrows point to the new forms. One space is empty. Can you decide what the procedure is and what should go in the empty space? State your reason. Other reasons may be possible. Can you find another?

$$3c + 5 = -13 \longrightarrow \quad -6$$

$$2m - 7 = 9 \longrightarrow \quad ?$$

_____

_____

_____

_____

_____

2.12

In the following diagram, two algebraic expressions are being changed or related to new forms following the same procedure or process. The arrows point to the new forms. One space is empty. Can you decide what the procedure is and what should go in the empty space? State your reason. Other reasons may be possible. Can you find another?

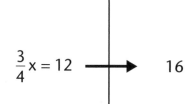

$$\frac{3}{4}x = 12 \longrightarrow 16$$

$$\frac{-3}{5}a = 6 \longrightarrow ?$$

NAME _____   DATE _____

 **2.13**

In the following diagram, two algebraic expressions are being changed or related to new forms following the same procedure or process. The arrows point to the new forms. One space is empty. Can you decide what the procedure is and what should go in the empty space? State your reason. Other reasons may be possible. Can you find another?

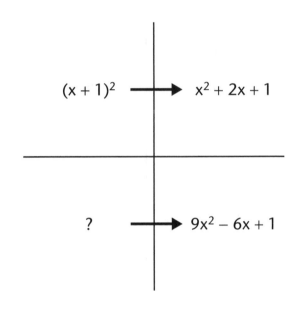

$(x + 1)^2 \longrightarrow x^2 + 2x + 1$

$? \longrightarrow 9x^2 - 6x + 1$

_____

_____

_____

_____

_____

2.14

In the following diagram, two algebraic expressions are being changed or related to new forms following the same procedure or process. The arrows point to the new forms. One space is empty. Can you decide what the procedure is and what should go in the empty space? State your reason. Other reasons may be possible. Can you find another?

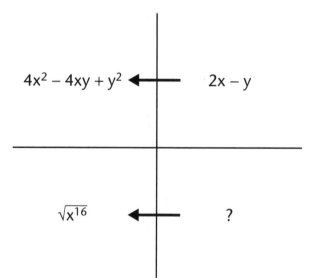

$4x^2 - 4xy + y^2$ ← $2x - y$

$\sqrt{x^{16}}$ ← ?

_____

_____

_____

_____

_____

2.15

In the following diagram, two algebraic expressions are being changed or related to new forms following the same procedure or process. The arrows point to the new forms. One space is empty. Can you decide what the procedure is and what should go in the empty space? State your reason. Other reasons may be possible. Can you find another?

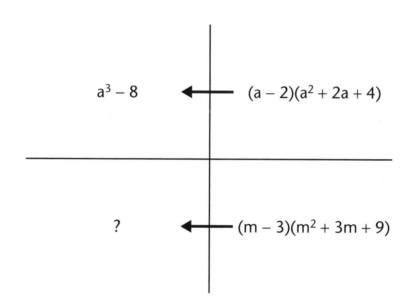

$a^3 - 8$ ⟵ $(a - 2)(a^2 + 2a + 4)$

? ⟵ $(m - 3)(m^2 + 3m + 9)$

_____

_____

_____

_____

_____

 **2.16**

In the following diagram, two algebraic expressions are being changed or related to new forms following the same procedure or process. The arrows point to the new forms. One space is empty. Can you decide what the procedure is and what should go in the empty space? State your reason. Other reasons may be possible. Can you find another?

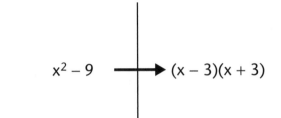

$$x^2 - 9 \longrightarrow (x - 3)(x + 3)$$

$$x^4 - 16 \longrightarrow \ ?$$

_____

_____

_____

_____

_____

2.17

In the following diagram, two algebraic expressions are being changed or related to new forms following the same procedure or process. The arrows point to the new forms. One space is empty. Can you decide what the procedure is and what should go in the empty space? State your reason. Other reasons may be possible. Can you find another?

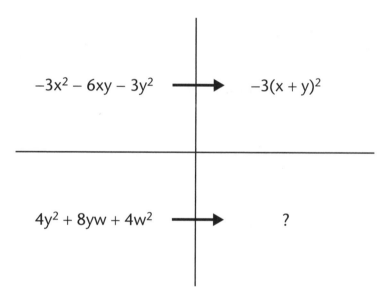

$-3x^2 - 6xy - 3y^2 \longrightarrow -3(x + y)^2$

$4y^2 + 8yw + 4w^2 \longrightarrow \quad ?$

_____

_____

_____

_____

_____

_____

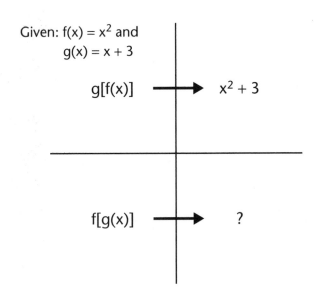 **2.18**

In the following diagram, two algebraic expressions are being changed or related to new forms following the same procedure or process. The arrows point to the new forms. One space is empty. Can you decide what the procedure is and what should go in the empty space? State your reason. Other reasons may be possible. Can you find another?

Given: $f(x) = x^2$ and
$g(x) = x + 3$

$g[f(x)] \longrightarrow x^2 + 3$

$f[g(x)] \longrightarrow ?$

Copyright © 2010 by John Wiley & Sons, Inc., *The Algebra Teacher's Activity-a-Day*

NAME _____   DATE _____

2.19

In the following diagram, two algebraic expressions are being changed or related to new forms following the same procedure or process. The arrows point to the new forms. One space is empty. Can you decide what the procedure is and what should go in the empty space? State your reason. Other reasons may be possible. Can you find another?

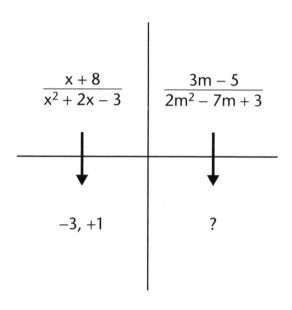

$$\frac{x + 8}{x^2 + 2x - 3} \qquad\qquad \frac{3m - 5}{2m^2 - 7m + 3}$$

−3, +1   ?

_____

_____

_____

_____

_____

2.20

In the following diagram, two algebraic expressions are being changed or related to new forms following the same procedure or process. The arrows point to the new forms. One space is empty. Can you decide what the procedure is and what should go in the empty space? State your reason. Other reasons may be possible. Can you find another?

$(4c + 5d)^2$ $\longrightarrow$ $16c^2 + 40cd + 25d^2$

$[(2a + 3b) + (4c + 5d)]^2 \longrightarrow$?

Copyright © 2010 by John Wiley & Sons, Inc., *The Algebra Teacher's Activity-a-Day*

SECTION THREE

Where Is It?

In these activities, students must use clues provided in a problem to eliminate items shown in eight of nine numbered boxes in a set or grid. The remaining item will then be the answer to the problem, because it satisfies all the clues. These problems require students to refine their understanding of mathematical definitions in order to correctly eliminate various items in the grid.

Example 3

In the following set of nine items, find the one item that satisfies all of the clues given.

➲ It has degree 9.

➲ It has two variables.

➲ It has a negative integral coefficient when simplified.

Where is it? (Indicate the box number of the correct answer.)

$27x^2y$ 1	$-9(y^3)^3$ 2	$9(xy)^3$ 3
$-3x^9$ 4	$(-3x^2y)^3$ 5	$12(xy)^2$ 6
$3x^5y^4$ 7	$-27x^6y$ 8	$-6xy^2$ 9

Explanation: The first clue requires a polynomial of degree 9; this eliminates the monomials in boxes *1*, *3*, *6*, *8*, and *9*. Students should cross out these boxes to show their elimination. Now only boxes *2*, *4*, *5*, and *7* remain. The second clue requires two variables, thereby eliminating the monomials in boxes *2* and *4*. Only boxes *5* and *7* remain. Finally, the third clue requires a negative integer for the coefficient; this eliminates box *7*, leaving the item in box *5* as the answer. The expression $(-3x^2y)^3$ simplifies to $-27x^6y^3$, which satisfies all three clues.

The answer key provided for this section identifies merely the box where the correct item is located. The process of elimination is the main problem-solving strategy used with this type of problem. When the class discusses each problem, have students clarify definitions of the terms included in the clues.

3.1

In the following set of nine items, find the one item that satisfies all of the clues given.

➲ It is a Pythagorean triple.

➲ All numbers are divisible by 3.

➲ Adjacent numbers differ by 3.

Where is it? (Indicate the box number of the correct answer.)

3, 4, 5	5, 10, 15	3, 6, 9
1	*2*	*3*
9, 15, 18	6, 8, 10	12, 18, 21
4	*5*	*6*
9, 12, 15	2, 5, 8	5, 12, 13
7	*8*	*9*

3.2

In the following set of nine items, find the one item that satisfies all of the clues given.

➲ It involves addition or subtraction.

➲ Its value is greater than −6.

➲ It is an inverse of an absolute value.

Where is it? (Indicate the box number of the correct answer.)

$\mid 0 \mid$ 1	$\mid 2 + 3 - 7 \mid$ 2	$-\mid 3 - (-2) \mid$ 3
$-\mid 3 - 9 \mid$ 4	$\mid -5 \mid$ 5	$(5)\mid -1 \mid$ 6
$\mid (-2) + (-3) \mid$ 7	$-\mid -5 \mid$ 8	$(-1)\mid +5 \mid$ 9

NAME _____   DATE _____

3.3

In the following set of nine items, find the one item that satisfies all of the clues given. To find x:

➲ Its absolute value is less than 1.

➲ x < −0.5

➲ It is not a mixed number.

Where is it? (Indicate the box number of the correct answer.)

$+\dfrac{2}{3}$ 1	$+3.6$ 2	$-\dfrac{1}{8}$ 3
-1.02 4	0 5	-0.05 6
$+0.7$ 7	$-\dfrac{3}{4}$ 8	$+1$ 9

3.4

In the following set of nine items, find the one item that satisfies all of the clues given.

➲ It is a monomial of degree 3.

➲ It has three variables.

➲ Its coefficient is a positive number.

Where is it? (Indicate the box number of the correct answer.)

x^2y 1	$8x^3$ 2	$-\dfrac{3}{xy}$ 3
$5xy^2w$ 4	$12xy^2$ 5	$6xyz$ 6
$-2yzw$ 7	$\dfrac{1}{xyz}$ 8	-18 9

3.5

In the following set of nine items, find the one item that satisfies all of the clues given.

➲ It has degree 9.

➲ It has two variables.

➲ It has a negative integral coefficient when simplified.

Where is it? (Indicate the box number of the correct answer.)

$27x^2y$	$-9(y^3)^3$	$9(xy)^3$
1	2	3
$-3x^9$	$(-3x^2y)^3$	$12(xy)^2$
4	5	6
$3x^5y^4$	$-27x^6y$	$-6xy^2$
7	8	9

3.6

In the following set of nine items, find the one item that satisfies all of the clues given.

➲ It is a line.

➲ It intersects the second quadrant.

➲ Its slope is greater than +1.

Where is it? (Indicate the box number of the correct answer.)

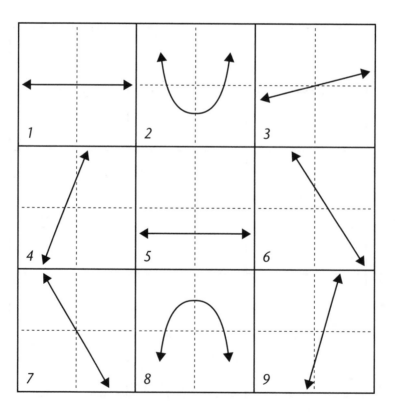

NAME _____   DATE _____

 3.7

In the following set of nine items, find the one item that satisfies all of the clues given.

➲ Its denominators are positive integers.

➲ One numerator is a monomial.

➲ Its solution is less than −1.

Where is it? (Indicate the box number of the correct answer.)

$\dfrac{3x}{8} = \dfrac{x-2}{3}$ 1	$\dfrac{3x}{-8} = \dfrac{x-2}{3}$ 2	$\dfrac{5x}{7} = \dfrac{-x+1}{-3}$ 3
$\dfrac{5x}{7} = \dfrac{x+1}{3}$ 4	$\dfrac{5x}{-7} = \dfrac{-x-1}{3}$ 5	$\dfrac{x-1}{5} = \dfrac{x+2}{4}$ 6
$\dfrac{3x}{8} = \dfrac{x+2}{3}$ 7	$\dfrac{x+1}{5} = \dfrac{-x+2}{-4}$ 8	$\dfrac{3x}{-8} = \dfrac{x+2}{3}$ 9

3.8

In the following set of nine items, find the one item that satisfies all of the clues given.

⮕ It is a linear inequality.

⮕ It has variables on both sides.

⮕ Its solution x is greater than +1.5.

Where is it? (Indicate the box number of the correct answer.)

$x^2 > x - 3$ 1	$x + 5 = 18$ 2	$3x + 5 < 7x - 1$ 3
$2x - 1 > 3x + 4$ 4	$3x^2 - x + 1 < 0$ 5	$8 > 5x - 1$ 6
$3x + 5 = 7x - 1$ 7	$2x - 3 < -9$ 8	$5 + 3x > 7x - 1$ 9

NAME _____   DATE _____

 3.9

In the following set of nine items, find the one item that satisfies all of the clues given. To simplify:

➲ Addition or subtraction must be done first.

➲ Multiplication by only 2x is needed last.

➲ Its exponent is applied as the second step.

Where is it? (Indicate the box number of the correct answer.)

$(2x)^3(x + 3)$ 1	$2x(3 + x - 5)^2$ 2	$3 + 2x - 1$ 3
$x(5 - x + 1)^2$ 4	$(x - 2)^3 + 6$ 5	$2x - (3)^2$ 6
$5(x + 2 + 1)^3$ 7	$(2x) + (7 - 1)^2$ 8	$(x + 7 - 3x)2x$ 9

3.10

In the following set of nine items, find the one item that satisfies all of the clues given.

➲ It is a trinomial.

➲ It has a factor of $(x + 1)$.

➲ Its linear coefficient is a negative integer.

Where is it? (Indicate the box number of the correct answer.)

$x^2 + 4$ 1	$x^2 - 3x + 2$ 2	$3x - 2$ 3
$x^2 - x - 2$ 4	$x^2 + 3x + 2$ 5	$x^2 + 4x + 3$ 6
$\dfrac{-5}{x + y - 4}$ 7	$x^2 - 4x + 3$ 8	$2x^2 + 2$ 9

NAME _____   DATE _____

 3.11

In the following set of nine items, find the one item that satisfies all of the clues given.

➲ It has a prime factor of $(x - 2)$.

➲ It is relatively prime to $x^2 - 7x + 6$.

➲ It has a prime factor of 3.

Where is it? (Indicate the box number of the correct answer.)

$2x^2 + 5x - 3$ 1	$3x^2 - 3$ 2	$x^2 - 7x + 6$ 3
$x^2 - 4$ 4	$3x^2 - 5x - 2$ 5	$x^2 - 4x + 4$ 6
$3x^2 - 9x + 6$ 7	$x^2 + 2x - 3$ 8	$3x^2 - 3x - 6$ 9

3.12

In the following set of nine items, find the one item that satisfies all of the clues given.

➲ It is completely simplified.

➲ It has three variables.

➲ It has b in the radicand.

Where is it? (Indicate the box number of the correct answer.)

$xa(\sqrt[3]{ab^2x})$	$\sqrt{a^2bx^3}$	$a\sqrt{bx^3}$
1	2	3
$\sqrt[3]{a^4b^2x^4}$	$\sqrt{bx}$	$a(\sqrt[3]{ab^2x^4})$
4	5	6
$b(\sqrt[5]{a^3x^4})$	$x(\sqrt[3]{a^4b^2x})$	$b\sqrt{ax}$
7	8	9

NAME _____ DATE _____

 3.13

In the following set of nine items, find the one item that satisfies all of the clues given.

➲ It has no transversal.

➲ It does not have exactly one intersection point.

➲ It consists of at least two lines.

Where is it? (Indicate the box number of the correct answer.)

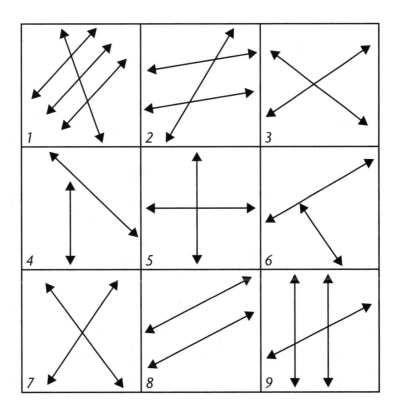

3.14

In the following set of nine items, find the one item that satisfies all of the clues given.

⮑ It is a parabola.

⮑ It does not pass through the origin.

⮑ As its x-values increase infinitely, its y-values decrease.

Where is it? (Indicate the box number of the correct answer.)

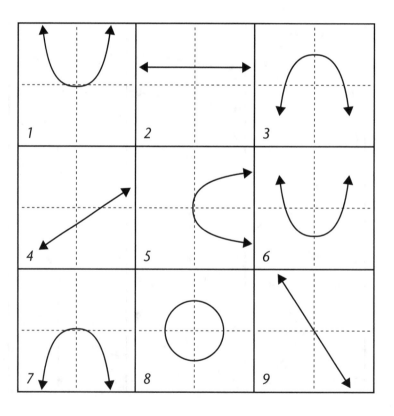

NAME _____   **DATE** _____

 3.15

In the following set of nine items, find the one item that satisfies all of the clues given.

⊃ A system consists of two linear equations.

⊃ It has one solution.

⊃ The graph of its solution lies in the fourth quadrant.

Where is the graph of the system? (Indicate the box number of the correct answer.)

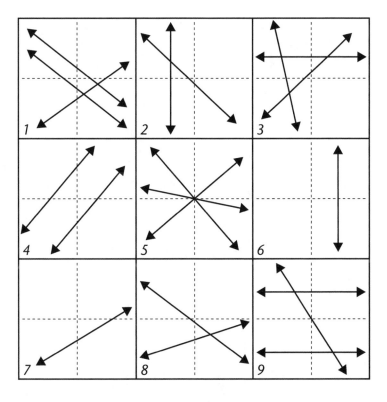

SECTION FOUR

Algebraic Pathways

In these activities, students must find one or more paths through boxes of a grid, following a logical order of steps needed to solve a given problem. Paths must always move "forward," that is, sideways, straight down, or diagonally downward, but never upward. These path "rules" are designed to prevent a student from *reversing* steps once a solution process has begun. For example, when solving $x - 3 = 4x + 6$, the goal might be to isolate the variable on the left side of the equation. So $+3$ might be *added* to both sides to obtain $x = 4x + 9$. If 9 is then *subtracted* from both sides to get $x - 9 = 4x$, the student has "undone" or reversed the previous addition step, which is an inefficient approach to solving the equation. If a path approaches a box that is not needed, the path should be drawn along the edges of the box. Emphasis is on the various procedures or sequences of steps that are possible for the same problem. Each path is a solution to the original problem. Algebraic pathways require logical reasoning and a careful analysis of solution steps.

Example 4

Find the answer to the following problem by drawing a path through the appropriate boxes in correct order. The path can move only sideways (left or right), straight down, or diagonally downward. It cannot move in an upward direction. To skip a box, draw along its edges. Try to find more than one path that works. Draw each new path in a different color. What steps might be done mentally?

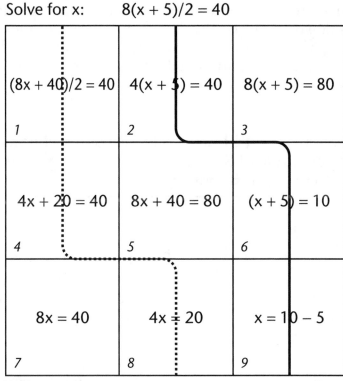

Solve for x: $8(x + 5)/2 = 40$

$(8x + 40)/2 = 40$	$4(x + 5) = 40$	$8(x + 5) = 80$
1	*2*	*3*
$4x + 20 = 40$	$8x + 40 = 80$	$(x + 5) = 10$
4	*5*	*6*
$8x = 40$	$4x = 20$	$x = 10 - 5$
7	*8*	*9*

Solution: $x = 5$

Explanation: This problem may be solved in several different ways. Students might decide to begin with box *2*, then divide by 4 to enter box *6*, and subtract 5 to enter box *9*. They finally reach the solution space and record $x = 5$. Another path would be through box *1* with a distribution of 8, followed by term divisions by 2 to enter box *4* and subtraction of 20 to enter box *8*. Division by 4 then leads to the solution space. Two other paths are possible (*3-5-7* and *3-6-9*). Each path should be drawn in a different color of pencil. Only the first two paths are represented in the example diagram. Students should be encouraged to find more than one possible path for a problem and during class discussion should give reasons for the paths they have taken.

4.1

Find the answer to the following problem by drawing a path through the appropriate boxes in correct order. The path can move only sideways (left or right), straight down, or diagonally downward. It cannot move in an upward direction. To skip a box, draw along its edges. Try to find more than one path that works. Draw each new path in a different color. What steps might be done mentally?

A right triangle has a leg length of 6 feet and a hypotenuse of 8 feet. Find the other leg length x to the nearest tenth of a foot.

$6^2 + 8^2 = x^2$ 1	$8^2 - 6^2 = x^2$ 2	$x^2 + 6^2 = 8^2$ 3
$64 - 36 = x^2$ 4	$x^2 + 36 = 64$ 5	$36 + 64 = x^2$ 6
$x^2 = 28$ 7	$100 = x^2$ 8	$x^2 = -28$ 9

Leg length x =

NAME _____   DATE _____

 **4.2**

Find the answer to the following problem by drawing a path through the appropriate boxes in correct order. The path can move only sideways (left or right), straight down, or diagonally downward. It cannot move in an upward direction. To skip a box, draw along its edges. Try to find more than one path that works. Draw each new path in a different color. What steps might be done mentally?

If the area of a triangle is 20 square units and the height is 8 units, what is the base b of the triangle in units?

½ b(8) = 20	½ (20)(8) = b	(8)(20) = ½ b
1	*2*	*3*
4b = 20	160 = ½ b	8b = 40
4	*5*	*6*
b = 2(160)	b = 20/4	b = 40/8
7	*8*	*9*

Base b =

4.3

Find the answer to the following problem by drawing a path through the appropriate boxes in correct order. The path can move only sideways (left or right), straight down, or diagonally downward. It cannot move in an upward direction. To skip a box, draw along its edges. Try to find more than one path that works. Draw each new path in a different color. What steps might be done mentally?

Express $(3^{-2})^3$ as a common fraction.

$\left(\dfrac{1}{3^2}\right)^3$ 1	$(3)^{-2 \cdot 3}$ 2	3^6 3
$\dfrac{1^3}{3^2}$ 4	$\dfrac{1}{(3^2)^3}$ 5	3^{-6} 6
$\dfrac{1}{3^2}$ 7	$\dfrac{1}{3^6}$ 8	$\dfrac{1}{3^{-6}}$ 9

Fraction =

NAME _____ **DATE** _____

 **4.4**

Find the answer to the following problem by drawing a path through the appropriate boxes in correct order. The path can move only sideways (left or right), straight down, or diagonally downward. It cannot move in an upward direction. To skip a box, draw along its edges. Try to find more than one path that works. Draw each new path in a different color. What steps might be done mentally?

Simplify and express result with a positive exponent: $(y^3)^5/y^2$.

$(y^3)^5 y^{-2}$ *1*	$\dfrac{y^8}{y^2}$ *2*	$\dfrac{y^{15}}{y^2}$ *3*
$(y^{15})^{-2}$ *4*	$(y^{15})(y^{-2})$ *5*	$(y^8)(y^{-2})$ *6*
y^{15-2} *7*	y^{-30} *8*	y^{8-2} *9*

Final form: _____

 **4.5**

Find the answer to the following problem by drawing a path through the
appropriate boxes in correct order. The path can move only sideways (left or
right), straight down, or diagonally downward. It cannot move in an upward
direction. To skip a box, draw along its edges. Try to find more than one path
that works. Draw each new path in a different color. What steps might be
done mentally?

Copyright © 2010 by John Wiley & Sons, Inc., *The Algebra Teacher's Activity-a-Day*

Simplify: $\dfrac{16a^2b^3}{5ac} \div \dfrac{8ab^2}{15c^2}$

$\dfrac{16ab^3(3c)}{5c(3c)} \div \dfrac{8ab^2}{15c^2}$ 1	$\dfrac{16ab^3}{5c} \div \dfrac{8ab^2}{15c^2}$ 2	$\dfrac{16ab^3}{5c} \cdot \dfrac{8ab^2}{15c^2}$ 3	$\dfrac{16a^2b^3}{5ac} \cdot \dfrac{15c^2}{8ab^2}$ 4
$\dfrac{48ab^3c}{15c^2} \div \dfrac{8ab^2}{15c^2}$ 5	$48ab^3c \div 8ab^2$ 6	$\dfrac{16a^2b^3}{a} \cdot \dfrac{3c}{8ab^2}$ 7	$\dfrac{128a^2b^5}{75c^3}$ 8
$\dfrac{6b^3c}{b^2}$ 9	$\dfrac{48ab^3c}{8ab^2}$ 10	$\dfrac{2ab}{a} \cdot \dfrac{3c}{1}$ 11	$(2b)(3c)$ 12

Final form:

NAME _____   DATE _____

 **4.6**

Find the answer to the following problem by drawing a path through the appropriate boxes in correct order. The path can move only sideways (left or right), straight down, or diagonally downward. It cannot move in an upward direction. To skip a box, draw along its edges. Try to find more than one path that works. Draw each new path in a different color. What steps might be done mentally?

Simplify: $2(3 + x) - 4x + 5(x - 1)$

$6 + 2x - 4x$ $+ 5(x - 1)$ 1	$6 + x - 4x$ $+ 5(x - 1)$ 2	$6 - 2x$ $+ 5(x - 1)$ 3
$6 - 3x + 5x - 1$ 4	$2(3 + x) - 4x$ $+ 5x - 5$ 5	$6 - 2x + 5x - 5$ 6
$(6 - 1)$ $+ (5x - 3x)$ 7	$(6 - 5)$ $+ (5x - 2x)$ 8	$6 + 2x + x - 5$ 9

Final form: _____

 **4.7**

Find the answer to the following problem by drawing a path through the appropriate boxes in correct order. The path can move only sideways (left or right), straight down, or diagonally downward. It cannot move in an upward direction. To skip a box, draw along its edges. Try to find more than one path that works. Draw each new path in a different color. What steps might be done mentally?

Solve for x: $3(2x - 5) = 33$

$6x - 15 = 33$ *1*	$6x - 5 = 33$ *2*	$2x - 5 = 11$ *3*
$6x = 33 + 15$ *4*	$2x = 11 + 5$ *5*	$6x = 33 + 5$ *6*
$2x = 16$ *7*	$6x = 48$ *8*	$6x = 38$ *9*

Solution: x =

Copyright © 2010 by John Wiley & Sons, Inc., *The Algebra Teacher's Activity-a-Day*

NAME _____   DATE _____

 **4.8**

Find the answer to the following problem by drawing a path through the appropriate boxes in correct order. The path can move only sideways (left or right), straight down, or diagonally downward. It cannot move in an upward direction. To skip a box, draw along its edges. Try to find more than one path that works. Draw each new path in a different color. What steps might be done mentally?

Solve for x: $2(x + 5) = 4(x - 3)$

$2x + 10 =$ $4x - 12$ *1*	$x + 5 =$ $2(x - 3)$ *2*	$2x + 5 =$ $4x - 3$ *3*
$x + 5 =$ $2x - 6$ *4*	$-4x + 2x =$ $-10 - 12$ *5*	$5 = 2x - 3$ *6*
$x = 2x - 11$ *7*	$+11 = 2x - x$ *8*	$-2x = -22$ *9*

Solution: x =

 **4.9**

Find the answer to the following problem by drawing a path through the appropriate boxes in correct order. The path can move only sideways (left or right), straight down, or diagonally downward. It cannot move in an upward direction. To skip a box, draw along its edges. Try to find more than one path that works. Draw each new path in a different color. What steps might be done mentally?

Solve for x: $8(x + 5)/2 = 40$

$(8x + 40)/2 = 40$ 1	$4(x + 5) = 40$ 2	$8(x + 5) = 80$ 3
$4x + 20 = 40$ 4	$8x + 40 = 80$ 5	$(x + 5) = 10$ 6
$8x = 40$ 7	$4x = 20$ 8	$x = 10 - 5$ 9

Solution: x =

Copyright © 2010 by John Wiley & Sons, Inc., *The Algebra Teacher's Activity-a-Day*

 **4.10**

Find the answer to the following problem by drawing a path through the appropriate boxes in correct order. The path can move only sideways (left or right), straight down, or diagonally downward. It cannot move in an upward direction. To skip a box, draw along its edges. Try to find more than one path that works. Draw each new path in a different color. What steps might be done mentally?

Find the solution set for x: $2x + 4 < 4x - 2$

$2x + 4 - 2x <$ $4x - 2 - 2x$ 1	$4 + 2 <$ $2x - 2 + 2$ 2	$2x + 4 - 4x <$ $4x - 2 - 4x$ 3
$6 < 2x$ 4	$-2x < -6$ 5	$-2x + 4 - 4 <$ $-2 - 4$ 6
$-x < -3$ 7	$\frac{1}{2}(6) < \frac{1}{2}(2x)$ 8	$-\frac{1}{2}(-2x) >$ $-\frac{1}{2}(-6)$ 9

Solution set for x:

Graph the solution set on a number line.

 **4.11**

Find the answer to the following problem by drawing a path through the appropriate boxes in correct order. The path can move only sideways (left or right), straight down, or diagonally downward. It cannot move in an upward direction. To skip a box, draw along its edges. Try to find more than one path that works. Draw each new path in a different color. What steps might be done mentally?

Find the solution set for x: $5x - 4 < 9x + 16$

$5x - 4 - 9x + 4 <$ $9x + 16 - 9x + 4$ 1	$5x - 4 - 16 <$ $9x$ 2	$-16 - 4 <$ $9x - 5x$ 3
$5x - 9x <$ $16 + 4$ 4	$-4x < 20$ 5	$(-4x)/4 <$ $(20/4)$ 6
$-20 < 4x$ 7	$(-20)/4 <$ $(4x)/4$ 8	$-x < 5$ 9

Solution set for x:

Graph the solution set on a number line.

NAME _____   DATE _____

4.12

Find the answer to the following problem by drawing a path through the appropriate boxes in correct order. The path can move only sideways (left or right), straight down, or diagonally downward. It cannot move in an upward direction. To skip a box, draw along its edges. Try to find more than one path that works. Draw each new path in a different color. What steps might be done mentally?

Solve for N: $\dfrac{N}{8} = \dfrac{49}{56}$

$8 \cdot 49 = 56N$ 1	$8\left(\dfrac{N}{8}\right) = 8\left(\dfrac{49}{56}\right)$ 2	$49N = 8 \cdot 56$ 3
$392 = 56N$ 4	$49N = 448$ 5	$1 \cdot N = \dfrac{8 \cdot 49}{56}$ 6
$N = \dfrac{448}{49}$ 7	$\dfrac{392}{56} = N$ 8	$N = \dfrac{49}{7}$ 9

Solution: N =

NAME _____ DATE _____

 **4.13**

Find the answer to the following problem by drawing a path through the appropriate boxes in correct order. The path can move only sideways (left or right), straight down, or diagonally downward. It cannot move in an upward direction. To skip a box, draw along its edges. Try to find more than one path that works. Draw each new path in a different color. What steps might be done mentally?

Find N: 20% of N = 40

$\dfrac{20}{100}N = 40$ 1	$\dfrac{40}{N} = \dfrac{20}{100}$ 2	$20N = 40$ 3	$\dfrac{N}{40} = \dfrac{20}{100}$ 4
$N = \dfrac{40}{20}$ 5	$\dfrac{1}{5}N = 40$ 6	$20N = 4000$ 7	$N = \dfrac{20 \cdot 40}{100}$ 8
$\dfrac{5}{1} \cdot \dfrac{1}{5}N = \dfrac{5}{1} \cdot 40$ 9	$N = 2$ 10	$N = \dfrac{800}{100}$ 11	$N = \dfrac{4000}{20}$ 12
$N = \dfrac{80}{10}$ 13	$N = 5 \cdot 40$ 14	$N = \dfrac{400}{2}$ 15	$N = \dfrac{400}{50}$ 16

Solution: N =

NAME _____ DATE _____

 **4.14**

Find the answer to the following problem by drawing a path through the appropriate boxes in correct order. The path can move only sideways (left or right), straight down, or diagonally downward. It cannot move in an upward direction. To skip a box, draw along its edges. Try to find more than one path that works. Draw each new path in a different color. What steps might be done mentally?

Solve for x: $3(x^2 + 2x + 1) = 27$

$3x^2 + 2x + 1 =$ 27 1	$3x^2 + 6x + 3 =$ 27 2	$x^2 + 2x + 1 =$ 24 3	$x^2 + 2x + 1 = 9$ 4
$3x^2 + 6x - 24 =$ 0 5	$3(x^2 + 2x - 8) =$ 0 6	$(x + 1)^2 = 9$ 7	$x^2 + 2x - 8 = 0$ 8
$(3x - 6)(x + 4) =$ 0 9	$x + 1 = +3$ or -3 10	$3(x + 4)(x - 2) =$ 0 11	$(x + 4)(x - 2) =$ 0 12
$x = -1 + 3$, or $x = -1 - 3$ 13	$3x - 6 = 0$, or $x + 4 = 0$ 14	$x + 4 = 0$, or $x - 2 = 0$ 15	$(3x + 12)(x - 2) =$ 0 16

Solution(s) for x:

4.15

Find the answer to the following problem by drawing a path through the appropriate boxes in correct order. The path can move only sideways (left or right), straight down, or diagonally downward. It cannot move in an upward direction. To skip a box, draw along its edges. Try to find more than one path that works. Draw each new path in a different color. What steps might be done mentally?

Solve for x: $(x + 3)^2 = 9$

$(x + 3)(x + 3) = 9$ *1*	$x + 3 = 81$ *2*	$x + 3 = \pm\sqrt{9}$ *3*
$x^2 + 6x + 9 = 9$ *4*	$x^2 + 6x = 0$ *5*	$x = 81 - 3$ *6*
$x(x + 6) = 0$ *7*	$x + 3 = \pm 3$ *8*	$x = 3 + 3$ *9*

Solution(s) for x:

Copyright © 2010 by John Wiley & Sons, Inc., *The Algebra Teacher's Activity-a-Day*

NAME _____ DATE _____

 **4.16**

Find the answer to the following problem by drawing a path through the appropriate boxes in correct order. The path can move only sideways (left or right), straight down, or diagonally downward. It cannot move in an upward direction. To skip a box, draw along its edges. Try to find more than one path that works. Draw each new path in a different color. What steps might be done mentally?

Solve for x: $(9/16)x^2 - 1 = 0$

$\left(\dfrac{9}{16}\right)x^2 = 1$ 1	$\left(\dfrac{3}{4}x - 1\right)\left(\dfrac{3}{4}x + 1\right) = 0$ 2	$x^2 - 1 = 0\left(\dfrac{16}{9}\right)$ 3	$x^2 = \dfrac{16}{9}$ 4
$x^2 = \dfrac{9}{16}$ 5	$(3x - 4)(3x + 4) = 0$ 6	$x^2 = 1$ 7	$\sqrt{x^2} = \pm\sqrt{\dfrac{16}{9}}$ 8
$x = \pm 1$ 9	$\dfrac{3}{4}x - 1 = 0$, or $\dfrac{3}{4}x + 1 = 0$ 10	$9x^2 = 16$ 11	$\dfrac{3}{4}x = +1$, or $\dfrac{3}{4}x = -1$ 12
$\sqrt{9x^2} = \pm\sqrt{16}$ 13	$3x = \pm 4$ 14	$x = +\dfrac{4}{3}, -\dfrac{4}{3}$ 15	$x = +\dfrac{3}{4}, -\dfrac{3}{4}$ 16

Solution(s) for x:

 **4.17**

Find the answer to the following problem by drawing a path through the appropriate boxes in correct order. The path can move only sideways (left or right), straight down, or diagonally downward. It cannot move in an upward direction. To skip a box, draw along its edges. Try to find more than one path that works. Draw each new path in a different color. What steps might be done mentally?

Solve for d: $-2(d + 6)^2 + 7 = -11$

$-2(d^2 + 12d + 36)$ $= -18$ 1	$-2(d^2 + 12d + 36)$ $+ 7 = -11$ 2	$-2(d^2 + 36) + 7$ $= -11$ 3	$-2(d + 6)^2 =$ -18 4
$-2d^2 - 24d - 72$ $+ 7 = -11$ 5	$d^2 + 12d + 36$ $= 9$ 6	$(d + 6)^2 = 9$ 7	$-2d^2 - 24d - 65$ $= -11$ 8
$d + 6 = \pm\sqrt{9}$ 9	$d^2 + 12d + 27$ $= 0$ 10	$-2(d^2 + 12d + 27)$ $= 0$ 11	$-2d^2 - 24d - 54$ $= 0$ 12
$(-2d - 6)(d + 9)$ $= 0$ 13	$d + 9 = 0$, or $d + 3 = 0$ 14	$(d + 9)(d + 3)$ $= 0$ 15	$d + 6 = 3$, or $d + 6 = -3$ 16

Solution(s) for d:

NAME _____ DATE _____

 4.18

Find the answer to the following problem by drawing a path through the appropriate boxes in correct order. The path can move only sideways (left or right), straight down, or diagonally downward. It cannot move in an upward direction. To skip a box, draw along its edges. Try to find more than one path that works. Draw each new path in a different color. What steps might be done mentally?

Solve for x: $2\sqrt{3x - 5} = 4$

$\sqrt{3x - 5} = 2$ *1*	$\sqrt{3x - 5} = 4$ *2*	$4(3x - 5) = 16$ *3*	$2(3x - 5) = 16$ *4*
$3x - 5 = 16$ *5*	$3x - 5 = 4$ *6*	$6x - 10 = 16$ *7*	$12x - 20 = 16$ *8*
$3x = 21$ *9*	$3x = 9$ *10*	$6x = 18$ *11*	$12x = 36$ *12*

Solution for x:

 **4.19**

Find the answer to the following problem by drawing a path through the appropriate boxes in correct order. The path can move only sideways (left or right), straight down, or diagonally downward. It cannot move in an upward direction. To skip a box, draw along its edges. Try to find more than one path that works. Draw each new path in a different color. What steps might be done mentally?

Find two consecutive odd positive integers the sum of whose squares is 74.

$N^2 + (N + 1)^2$ $= 74$ *1*	$2N^2 + 2N + 1$ $= 74$ *2*	$N^2 + (N + 2)^2$ $= 74$ *3*
$(2N - 10)(N + 7)$ $= 0$ *4*	$2N^2 + 4N - 70$ $= 0$ *5*	$2N^2 + 2N - 73$ $= 0$ *6*
$2(N - 5)(N + 7)$ $= 0$ *7*	$(N - 5)(N + 7)$ $= 0$ *8*	$N^2 + 2N - 35$ $= 0$ *9*

Solution(s) for N are:

Thus, the positive integers needed are:

NAME _____ DATE _____

 **4.20**

Find the answer to the following problem by drawing a path through the appropriate boxes in correct order. The path can move only sideways (left or right), straight down, or diagonally downward. It cannot move in an upward direction. To skip a box, draw along its edges. Try to find more than one path that works. Draw each new path in a different color. What steps might be done mentally?

Solve the system for x and y: $x - 2y = -3$ and $3x - 2y = -5$

$-3x + 6y = +9$ $3x - 2y = -5$ 1	$x - 2y = -3$ $-3x + 2y = -5$ 2	$x - 2y = -3$ $3x + 2y = +5$ 3	$x - 2y = -3$ $-3x + 2y = +5$ 4
$4y = +4$ 5	$(1/4)(4y) =$ $(1/4)(4)$ 6	$4x = +2$ 7	$-2x = +2$ 8
$-2x = -8$ 9	$(1/4)(4x) =$ $(1/4)(2)$ 10	$-x = +1$ 11	$(1/2)(-2x) =$ $(1/2)(2)$ 12
$y = +1$ 13	$x - 2(1) = -3$ 14	$(-1) - 2y = -3,$ or $-2y = -2$ 15	$x = -1$ 16

Solution: x = _____ and y = _____

(x, y) = (_____ , _____)

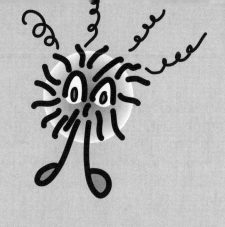

SECTION FIVE

Squiggles

In the activities in this section, students must form pairs of adjacent expressions according to given rules or relationships. The overall assignment of a given set of expressions to points on a "squiggle," or network, forms a solution for that squiggle. Other assignments of the same expressions on a network are possible. For some squiggles, one expression may be initially assigned to a point; other expressions provided must then be assigned to the remaining points according to the given rule. Students should be encouraged to search for a general strategy for assigning expressions to a particular squiggle—for example, Which expression might be paired with more than one other expression? That expression should then be assigned to a point belonging to several paths of the network.

Example 5

A rule is provided with this squiggle or network that tells you how expressions assigned to adjacent or connected points should relate to each other. Each expression or number must be assigned to only one point on the squiggle. Assign all algebraic expressions from the given set. A complete assignment of expressions to all the points represents a solution. More than one solution may be possible.

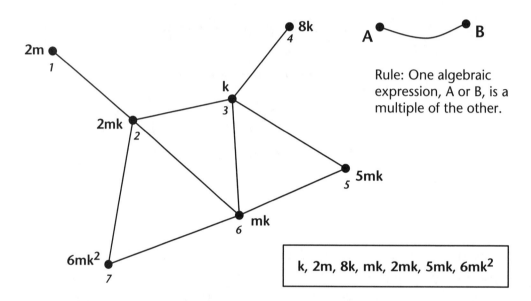

Rule: One algebraic expression, A or B, is a multiple of the other.

$$k, \ 2m, \ 8k, \ mk, \ 2mk, \ 5mk, \ 6mk^2$$

Explanation: Students should notice that if one expression in the box contains all factors of another expression, plus the first expression has an extra factor, then that first expression is a multiple of the second expression. One strategy is to assign the simplest expression to a point with several paths connected to that point. For example, assign the expression **k** to (3). Then **8k** might be assigned to (4), **5mk** to (5), **mk** to (6), and **2mk** to (2). This leaves **2m** and **6mk^2** to be assigned. **2mk** is a multiple of **2m**, but **2m** and **mk** are not multiples of each other; therefore, **2m** may be assigned to (1) but not to (7). **6mk^2** is a multiple of **mk** and also **2mk**, so it may be assigned to (7), which is connected to (2) and (6). Now each pair of connected points represents two expressions where one is a multiple of the other. This completed assignment forms a solution to the problem. In the answer key this solution is shown as follows: (1) 2m, (2) 2mk, (3) k, (4) 8k, (5) 5mk, (6) mk, (7) 6mk^2. A different assignment of expressions may be written on the same diagram in another color of pen or pencil so that students can easily see and discuss the new solution.

5.1

A rule is provided with this squiggle or network that tells you how expressions assigned to adjacent or connected points should relate to each other. Each expression or number must be assigned to only one point on the squiggle. Assign all numbers from the given set. One number has already been assigned. A complete assignment of expressions to all the points represents a solution. More than one solution may be possible.

Rule: The sum of integers A and B is prime and each prime sum occurs only once on the squiggle.

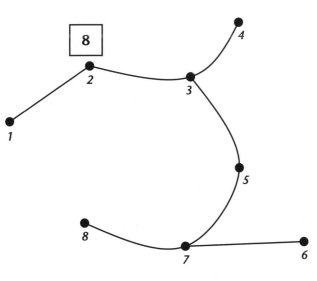

8

−25, −4, −2, 2, 5, 9, 27

5.2

A rule is provided with this squiggle or network that tells you how expressions assigned to adjacent or connected points should relate to each other. Each expression or number must be assigned to only one point on the squiggle. Assign all numbers from the given set. A complete assignment of expressions to all the points represents a solution. More than one solution may be possible.

Rule: For numbers A and B, one number is rational and the other is irrational.

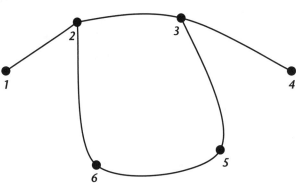

$$\sqrt{36}, \sqrt{2}, -3/4, \pi, 1.783129\ldots, -9$$

Copyright © 2010 by John Wiley & Sons, Inc., *The Algebra Teacher's Activity-a-Day*

5.3

A rule is provided with this squiggle or network that tells you how expressions assigned to adjacent or connected points should relate to each other. Each expression or number must be assigned to only one point on the squiggle. Select any numbers that will satisfy the rule. One number has already been assigned. A complete assignment of expressions to all the points represents a solution. More than one solution may be possible.

Rule: One number, A or B, is 5 plus a multiple of the other number.

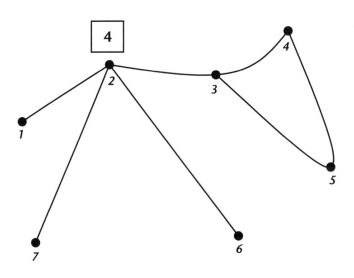

5.4

A rule is provided with this squiggle or network that tells you how expressions assigned to adjacent or connected points should relate to each other. Each expression or number must be assigned to only one point on the squiggle. Assign all numbers from the given set. One number has already been assigned. A complete assignment of expressions to all the points represents a solution. More than one solution may be possible.

A •⌣• B

Rule: Only one of square roots A and B can be simplified.

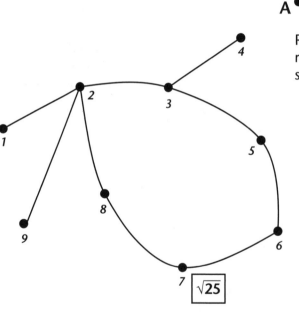

$$\sqrt{13}, \sqrt{27}, \sqrt{30}, \sqrt{51}, \sqrt{61}, \sqrt{72}, \sqrt{83}, \sqrt{99}$$

Copyright © 2010 by John Wiley & Sons, Inc., *The Algebra Teacher's Activity-a-Day*

NAME _____   DATE _____

5.5

A rule is provided with this squiggle or network that tells you how expressions assigned to adjacent or connected points should relate to each other. Each expression or number must be assigned to only one point on the squiggle. Assign all algebraic expressions from the given set. A complete assignment of expressions to all the points represents a solution. More than one solution may be possible.

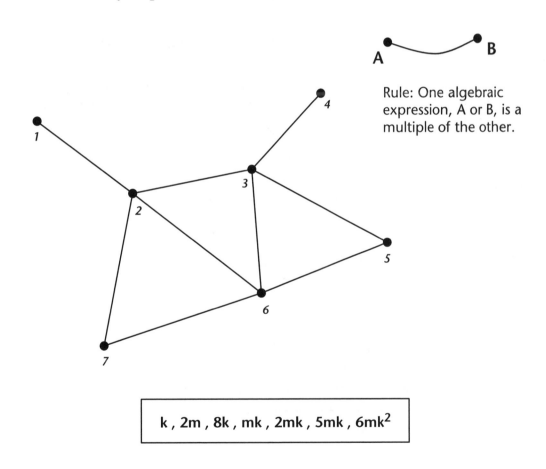

Rule: One algebraic expression, A or B, is a multiple of the other.

$$k , 2m , 8k , mk , 2mk , 5mk , 6mk^2$$

5.6

A rule is provided with this squiggle or network that tells you how expressions assigned to adjacent or connected points should relate to each other. Each expression or number must be assigned to only one point on the squiggle. Assign all expressions from the given set. A complete assignment of expressions to all the points represents a solution. More than one solution may be possible.

Rule: The difference between A and B has a factor of 3.

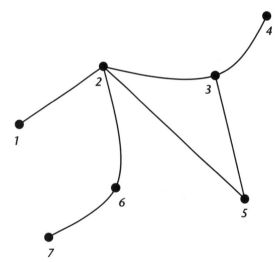

-9m, -6m, -3m, 3m, 6m, 9m, 12m

5.7

A rule is provided with this squiggle or network that tells you how expressions assigned to adjacent or connected points should relate to each other. Each expression or number must be assigned to only one point on the squiggle. Assign all expressions from the given set. One expression has already been assigned. A complete assignment of expressions to all the points represents a solution. More than one solution may be possible.

A •⌣⌣• B

Rule: Only one of the monomials, A and B, is of even degree; the other is of odd degree.

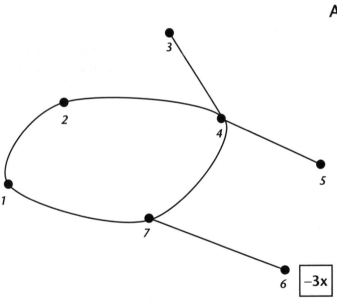

−3x

$5x^2y$, $8x^3w^2$, 12, ab^2c, $(mb)/2$, $3w^2$

5.8

A rule is provided with this squiggle or network that tells you how expressions assigned to adjacent or connected points should relate to each other. Each expression or number must be assigned to only one point on the squiggle. Assign all expressions from the given set. A complete assignment of expressions to all the points represents a solution. More than one solution may be possible.

A •⁀• B

Rule: One expression, A or B, is equal to the other expression to some power.

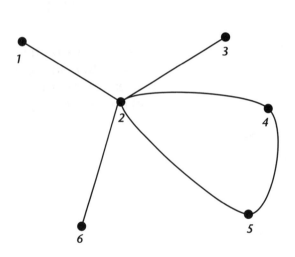

$$1,\ 2\sqrt{x},\ 4x,\ 64x^3,\ (4x)^{1/3},\ 16x^2$$

NAME _____   DATE _____

5.9

A rule is provided with this squiggle or network that tells you how expressions assigned to adjacent or connected points should relate to each other. Each expression or number must be assigned to only one point on the squiggle. Assign all expressions or formulas from the given list. A complete assignment of expressions to all the points represents a solution. More than one solution may be possible.

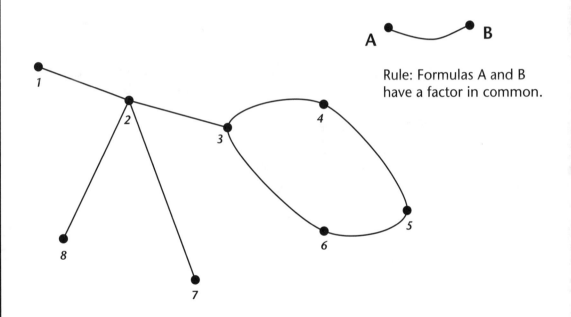

Rule: Formulas A and B have a factor in common.

Cone's volume $= (1/3)\pi r^2 h$ Circle's area $= \pi r^2$

Circumference $= 2\pi r$ Parallelogram's area $= bh$

Cylinder's volume $= \pi r^2 h$ Trapezoid's area $= (1/2)h(B + b)$

Triangle's area $= (bh)/2$ Rectangular prism's volume $= lwh$

5.10

A rule is provided with this squiggle or network that tells you how expressions assigned to adjacent or connected points should relate to each other. Each expression or number must be assigned to only one point on the squiggle. Assign all equations from the given set. A complete assignment of expressions to all the points represents a solution. More than one solution may be possible.

A   B

Rule: One of the graphs of A and B has the greater slope.

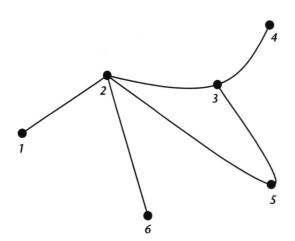

$y = 3x + 4$	$-3x + 4y - 20 = 0$	$2y + x = -2$
$y + 2x - 3 = 0$	$4y = 3x + 8$	$3x - y = 1$

Copyright © 2010 by John Wiley & Sons, Inc., *The Algebra Teacher's Activity-a-Day*

5.11

A rule is provided with this squiggle or network that tells you how expressions assigned to adjacent or connected points should relate to each other. Each expression or number must be assigned to only one point on the squiggle. Assign all equations from the given set. A complete assignment of expressions to all the points represents a solution. More than one solution may be possible.

Rule: The graphs of A and B have the same slope or the same y-intercept, but not both.

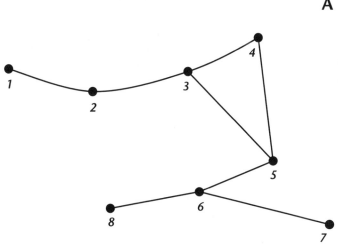

| $y = 4x - 5$ | $4x - y = -7$ | $y = x + 7$ | $x + 3y = 21$ |
| $2x + 6y = -30$ | $6y = -2x - 3$ | $y = x - 5$ | $8x - 2y = 1$ |

5.12

A rule is provided with this squiggle or network that tells you how expressions assigned to adjacent or connected points should relate to each other. Each expression or number must be assigned to only one point on the squiggle. Assign all expressions from the given set. One expression has already been assigned. A complete assignment of expressions to all the points represents a solution. More than one solution may be possible.

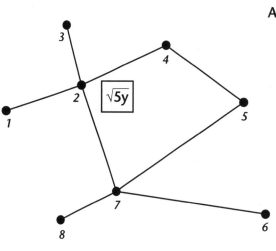

A •⌣• B

Rule: Only one of the expressions A and B can be simplified.

$\sqrt{5y}$

$$\sqrt[3]{8x^6},\ \sqrt{9m^2},\ \sqrt[3]{10x^2},\ (27y^9)^{1/3},\ \sqrt{13xy},\ \sqrt[4]{16b^8},\ \sqrt{6abc}$$

NAME _____   **DATE** _____

5.13

A rule is provided with this squiggle or network that tells you how expressions assigned to adjacent or connected points should relate to each other. Each expression or number must be assigned to only one point on the squiggle. Assign all expressions from the given set. A complete assignment of expressions to all the points represents a solution. More than one solution may be possible.

Rule: One expression, A or B, has more factors in its total number of factors than the other expression has.

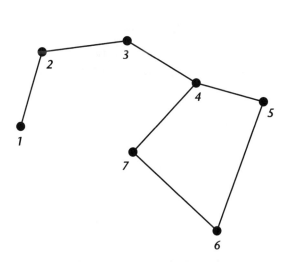

| 25, 12, 6y, 5x, 18, 2xy, 3x² |

5.14

A rule is provided with this squiggle or network that tells you how expressions assigned to adjacent or connected points should relate to each other. Each expression or number must be assigned to only one point on the squiggle. Assign all expressions from the given set. One expression has already been assigned. A complete assignment of expressions to all the points represents a solution. More than one solution may be possible.

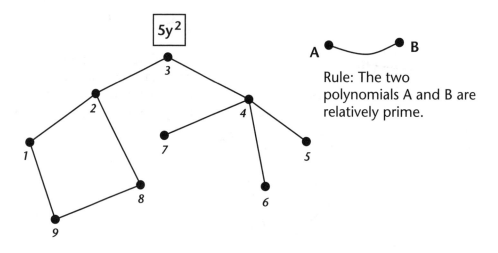

$5y^2$

A •⌣• B

Rule: The two polynomials A and B are relatively prime.

$x^2 + 2x - 3$	$3x^2 + 6x$	$2x + 6$	$5y$
$x^2 + x - 6$	$2x - 2$	$2xy - 2y$	$3x$

NAME _____   DATE _____

5.15

A rule is provided with this squiggle or network that tells you how expressions assigned to adjacent or connected points should relate to each other. Each expression or number must be assigned to only one point on the squiggle. Assign all expressions from the given set. One expression has already been assigned. A complete assignment of expressions to all the points represents a solution. More than one solution may be possible.

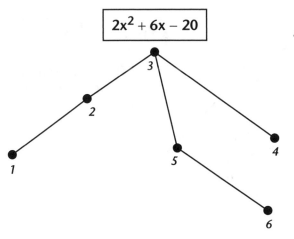

A •⌣• B

Rule: One expression, A or B, is a factor of the other.

$2, 2x - 4, x + 5, 2x + 10, x^2 + 3x - 10$

5.16

A rule is provided with this squiggle or network that tells you how expressions assigned to adjacent or connected points should relate to each other. Each expression or number must be assigned to only one point on the squiggle. Assign all expressions from the given set. One expression has already been assigned. A complete assignment of expressions to all the points represents a solution. More than one solution may be possible.

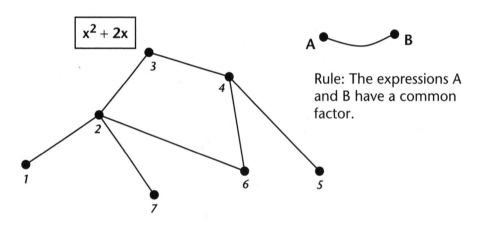

$x^2 + 2x$

A ●⌣● B

Rule: The expressions A and B have a common factor.

$$4x^2 - 16,\ 5x^2 - 15x,\ 9x^2,\ 3x^2 + 6x,\ 4x - 12,\ x^2 - x - 6$$

NAME _____   **DATE** _____

5.17

A rule is provided with this squiggle or network that tells you how expressions assigned to adjacent or connected points should relate to each other. Each expression or number must be assigned to only one point on the squiggle. Assign all expressions from the given set. A complete assignment of expressions to all the points represents a solution. More than one solution may be possible.

Rule: Polynomials A and B share only the common factor $(d + 1)$ or $(d + 2)$ or $(d - 3)$.

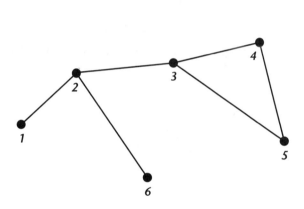

$2d^2 - 6d - 20$	$d^2 - d - 6$	$2d^2 + 8d + 8$
$d^2 - 5d + 6$	$d^2 - 2d - 3$	$d^2 + 3d + 2$

5.18

A rule is provided with this squiggle or network that tells you how expressions assigned to adjacent or connected points should relate to each other. Each expression or number must be assigned to only one point on the squiggle. Assign all expressions from the given set. One expression has already been assigned. A complete assignment of expressions to all the points represents a solution. More than one solution may be possible.

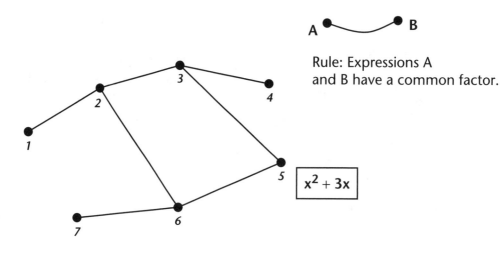

A ● B

Rule: Expressions A and B have a common factor.

$x^2 + 3x$

$x + 3$	$x^2 + x - 6$	$x^2 - 2x - 15$
$5x - 5$	$4x^2 - 16$	$x^2 + 2x - 3$

NAME _____ **DATE** _____

5.19

A rule is provided with this squiggle or network that tells you how expressions assigned to adjacent or connected points should relate to each other. Each expression or number must be assigned to only one point on the squiggle. Assign all equations from the given set. A complete assignment of expressions to all the points represents a solution. More than one solution may be possible.

Rule: The graphs of A and B have the same vertex or the same roots.

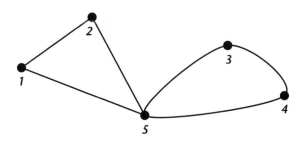

$$y = x^2 - 4 \qquad y = x^2 + 4 \qquad y = -x^2 + 4$$

$$y = (1/2)x^2 - 2 \qquad y = (-1/4)x^2 + 4$$

5.20

A rule is provided with this squiggle or network that tells you how expressions assigned to adjacent or connected points should relate to each other. Each expression or number must be assigned to only one point on the squiggle. Assign all equations from the given set. A complete assignment of expressions to all the points represents a solution. More than one solution may be possible.

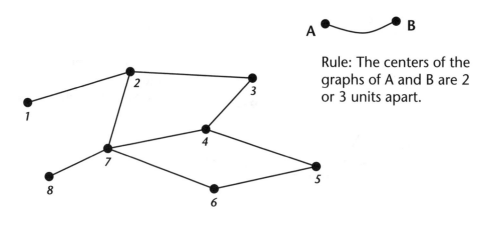

Rule: The centers of the graphs of A and B are 2 or 3 units apart.

$$x^2 + y^2 = 4 \qquad\qquad (x - 2)^2 + y^2 = 9$$

$$(x + 2)^2 + (y - 3)^2 = 1 \qquad\qquad x^2 + y^2 + 4y + 1 = 0$$

$$x^2 + y^2 - 4x + 4y + 7 = 0 \qquad\qquad (x - 2)^2 + (y - 3)^2 = 2$$

$$x^2 + y^2 - 6y + 4 = 0 \qquad\qquad (x + 3)^2 + y^2 = 4$$

Copyright © 2010 by John Wiley & Sons, Inc., *The Algebra Teacher's Activity-a-Day*

SECTION SIX

Math Mystery Messages

In the activities in this section, students must use partial codes to discover secret messages. Each message has a mathematical theme and is a complete sentence. Students must apply logical reasoning in order to speculate about missing letters. This requires drawing on their familiarity with the basic structure of the English language and with common mathematical ideas. Once each message is completed, students must give an example of the property or situation described in the message.

To decode a message, students should first record all given letters in their appropriately numbered spaces. Each letter has its own number throughout the message. Then students should notice the first word of the message. If a three-letter word is used, it is most likely the article *the*; however, common first words might also be *a* or *an*.

Two-letter words within the message might be prepositions such as *at*, *in*, or *of*; they might also be the verb *is*. Each time a new letter is found, all numbered spaces for that letter should be filled in. Then new words may be identified in the message, followed by more letters. Students should continue this process until the entire message has been determined.

Example 6

Use the numbers written below the spaces to discover the mystery message about a math idea. Each number represents a different letter of the alphabet. Some numbers are already shown with their respective letters. When you've decoded the message, write an example below the message to illustrate the math idea.

$\frac{\quad}{7}$ $\frac{\quad}{8}$ $\frac{E}{2}$ $\frac{\quad}{5}$ $\frac{\quad}{9}$ $\frac{\quad}{2}$ $\frac{\quad}{10}$ $\frac{\quad}{7}$ $\frac{\quad}{5}$ $\frac{\quad}{7}$ $\frac{\quad}{11}$

$\frac{\quad}{4}$ $\frac{\quad}{12}$ $\frac{\quad}{13}$ $\frac{\quad}{9}$ $\frac{\quad}{9}$ $\frac{\quad}{5}$ $\frac{\quad}{7}$ $\frac{\quad}{5}$ $\frac{\quad}{4}$ $\frac{\quad}{10}$

$\frac{\quad}{5}$ $\frac{\quad}{6}$ $\frac{Z}{1}$ $\frac{\quad}{2}$ $\frac{\quad}{3}$ $\frac{\quad}{4}$.

Example: _____

Explanation: Here is a possible reasoning process. When all known letters Z and E have been recorded in their appropriate spaces, only four spaces are filled. To find new letters, we might look at the last word of the sentence: Z E _ _. Because the message is a math idea, ZERO is a likely choice. Then R will fill the 3-space and O will fill the 4-space. Fill in all possible R's and O's.

Now look at the first word: _ _ E. In English, a common first word of a sentence is THE. Assign T to the 7-space and H to the 8-space. Fill in all possible T's and H's. Several spaces are still empty. Because ZERO is a noun, the second word from the end is probably OF or IS. Students may need to test each word to see which one provides the most new information. However, O is already in the 4-space, so O cannot be in the 5-space too. Thus we will use IS: I in the 5-space and S in the 6-space.

Because this word is the verb IS, the other two-letter word must be the preposition OF; F goes in the 12-space. Now we must apply what we know about the properties of ZERO to find the last missing letters. This leads us to the words IDENTITY and ADDITION. They correctly fill the empty spaces and complete the sentence.

T	H	E		I	D	E	N	T	I	T	Y
7	8	2		5	9	2	10	7	5	7	11

O	F		A	D	D	I	T	I	O	N
4	12		13	9	9	5	7	5	4	10

I	S		Z	E	R	O .
5	6		1	2	3	4

Here is a possible example to illustrate the mathematical idea in the message:

Example: $5 + 0 = 5$ and $0 + 5 = 5$

6.1

Use the numbers written below the spaces to discover the mystery message about a math idea. Each number represents a different letter of the alphabet. Some numbers are already shown with their respective letters. When you've decoded the message, write an example below the message to illustrate the math idea.

_	_	E		_	_	_	_	_	_	_	_
7	8	2		5	9	2	10	7	5	7	11

_	_		_	_	_	_	_	_	_	_
4	12		13	9	9	5	7	5	4	10

_	_	Z	_	_	_
5	6	1	2	3	4

Example: _____

Copyright © 2010 by John Wiley & Sons, Inc., *The Algebra Teacher's Activity-a-Day*

6.2

Use the numbers written below the spaces to discover the mystery message about a math idea. Each number represents a different letter of the alphabet. Some numbers are already shown with their respective letters. When you've decoded the message, write an example below the message to illustrate the math idea.

$\overline{}\ \overline{}\ \overline{}\quad \overline{}\ \overline{}\ \overline{}\ \overline{}\ \overline{}\ \overline{}\ \overline{}\ \overline{}$
1   2   3   4   5   1   6   7   6   8   9

$\overline{}\ \overline{}\ \overline{}\ \overset{T}{\overline{}}\quad \overline{}\ \overline{}\ \overset{E}{\overline{}}$
1   10   11   1   11   8   6

$\overline{}\ \overline{}\ \overline{}\ \overline{}\ \overline{}\ \overline{}\ \overline{}\ \overline{}\ \overset{S}{\overline{}}$
3   12   12   3   9   4   1   6   9

$\overline{}\ \overline{}\ \overline{}\quad \overline{}\ \overline{}\quad \overline{}\ \overline{}\ \overline{}\ \overset{O}{\overline{}}$.
9   13   14   1   3   15   6   8   3

Example: _____

6.3

Use the numbers written below the spaces to discover the mystery message about a math idea. Each number represents a different letter of the alphabet. Some numbers are already shown with their respective letters. When you've decoded the message, write an example below the message to illustrate the math idea.

$\dfrac{}{1}$ $\dfrac{N}{2}$ $\dfrac{}{3}$

$\dfrac{}{1}$ $\dfrac{}{4}$ $\dfrac{}{5}$ $\dfrac{}{6}$ $\dfrac{B}{7}$ $\dfrac{}{8}$ $\dfrac{}{1}$ $\dfrac{}{9}$ $\dfrac{}{10}$

$\dfrac{}{11}$ $\dfrac{}{6}$ $\dfrac{}{8}$ $\dfrac{}{12}$ $\dfrac{}{13}$ $\dfrac{}{14}$ $\dfrac{}{6}$ $\dfrac{}{8}$

$\dfrac{}{9}$ $\dfrac{}{11}$ $\dfrac{S}{15}$ $\dfrac{}{6}$ $\dfrac{L}{4}$ $\dfrac{}{16}$

$\dfrac{}{6}$ $\dfrac{}{17}$ $\dfrac{}{18}$ $\dfrac{A}{1}$ $\dfrac{}{4}$ $\dfrac{}{15}$

$\dfrac{T}{11}$ $\dfrac{}{19}$ $\dfrac{}{6}$ $\dfrac{}{9}$ $\dfrac{}{20}$ $\dfrac{}{6}$ $\dfrac{}{2}$ $\dfrac{}{11}$ $\dfrac{I}{9}$ $\dfrac{}{11}$ $\dfrac{}{3}$,

$\dfrac{}{13}$ $\dfrac{}{2}$ $\dfrac{}{6}$.

Example: _____

Copyright © 2010 by John Wiley & Sons, Inc., *The Algebra Teacher's Activity-a-Day*

6.4

Use the numbers written below the spaces to discover the mystery message about a math idea. Each number represents a different letter of the alphabet. Some numbers are already shown with their respective letters. When you've decoded the message, write an example below the message to illustrate the math idea.

$\overline{}_{13}$ $\overline{}_{9}$ $\overline{}_{4}$ $\overset{E}{\overline{}}_{3}$ $\overline{}_{5}$ $\overline{}_{6}$ $\overline{}_{13}$ $\overline{}_{17}$ $\overline{}_{15}$

$\overset{F}{\overline{}}_{9}$ $\overline{}_{15}$ $\overline{}_{14}$ $\overline{}_{1}$ $\overline{}_{6}$ $\overline{}_{5}$, $\overline{}_{1}$ $\overline{}_{2}$ $\overline{}_{3}$

$\overline{}_{7}$ $\overline{}_{5}$ $\overline{}_{6}$ $\overline{}_{18}$ $\overline{}_{11}$ $\overline{}_{14}$ $\overline{}_{1}$ $\overline{}_{13}$ $\overline{}_{17}$

$\overline{}_{15}$ $\overline{}_{12}$ $\overline{}_{19}$ $\overline{}_{15}$ $\overline{}_{8}$ $\overset{S}{\overline{}}_{17}$ $\overline{}_{4}$ $\overline{}_{3}$ $\overline{}_{5}$ $\overline{}_{6}$.

Example: _____

6.5

Use the numbers written below the spaces to discover the mystery message about a math idea. Each number represents a different letter of the alphabet. Some numbers are already shown with their respective letters. When you've decoded the message, write an example below the message to illustrate the math idea.

__ __ S__ __ __ __ __ __
4 5 6 7 8 9 1 3

__ __ __ U__ __   __ __   __
10 4 8 9 3   15 6   4

__ __ __ B__ __ __ ' __
12 9 13 5 3 14   6

__ __ __ T__ __ __ __ __
16 15 6 1 4 12 17 3

__ __ __ M__   Z__ __ __ O__ .
11 14 7 13   18 3 14 7

Example: _____

NAME _____ **DATE** _____

 6.6

Use the numbers written below the spaces to discover the mystery message about a math idea. Each number represents a different letter of the alphabet. Some numbers are already shown with their respective letters. When you've decoded the message, write an example below the message to illustrate the math idea.

$$\frac{}{1} \quad \frac{}{2} \quad \frac{}{3} \quad \frac{U}{4} \quad \frac{}{1} \quad \frac{}{5} \quad \frac{}{6}$$

$$\frac{}{5} \quad \frac{O}{7} \quad \frac{}{7} \quad \frac{}{8} \quad \quad \frac{}{9} \quad \frac{}{1} \quad \frac{Y}{10}$$

$$\frac{}{11} \quad \frac{}{1} \quad \frac{V}{12} \quad \frac{}{6} \quad \quad \frac{}{1}$$

$$\frac{}{13} \quad \frac{}{6} \quad \frac{}{14} \quad \frac{}{1} \quad \frac{}{8} \quad \frac{}{15} \quad \frac{}{12} \quad \frac{}{6}$$

$$\frac{}{12} \quad \frac{}{1} \quad \frac{}{16} \quad \frac{}{4} \quad \frac{}{6}.$$

Example: _____

6.7

Use the numbers written below the spaces to discover the mystery message about a math idea. Each number represents a different letter of the alphabet. Some numbers are already shown with their respective letters. When you've decoded the message, write an example below the message to illustrate the math idea.

						T		
6	16	3		12	15	1	2	3

		O				
4	5	6	7	8	9	1

6	10	1	11	6

R										
5	3	9	12	4	5	6	9	13	14	15

Example: _____

6.8

Use the numbers written below the spaces to discover the mystery message about a math idea. Each number represents a different letter of the alphabet. Some numbers are already shown with their respective letters. When you've decoded the message, write an example below the message to illustrate the math idea.

```
___   ___   ___   _N_
 1     2     3     4

_M_   ___   ___   _T_   ___   ___   ___   ___   ___   ___   ___ ,
 8     7     9    10    11    12     9     5    11     4    15

_C_   ___   ___   ___   ___   ___         ___   ___   ___
13     2    14     4    15     3          10     2     3

_O_   ___   ___   ___   ___         ___   ___         ___   ___   ___
 6    16    17     3    16           6    18          10     2     3

___   ___   ___   ___   ___   ___   ___         ___   ___
18    14    13    10     6    16    20          10     6

___   ___   ___         ___   ___   _E_         ___   ___   ___   ___
15     3    10          10     2     3          20    14     8     3

_P_   ___   ___   ___   ___   ___   ___ .
12    16     6    17     7    13    10
```

Example: _____

6.9

Use the numbers written below the spaces to discover the mystery message about a math idea. Each number represents a different letter of the alphabet. Some numbers are already shown with their respective letters. When you've decoded the message, write an example below the message to illustrate the math idea.

$\dfrac{E}{1}$ $\dfrac{X}{2}$ $\dfrac{}{3}$ $\dfrac{}{4}$ $\dfrac{}{5}$ $\dfrac{}{1}$ $\dfrac{}{5}$ $\dfrac{}{6}$ $\dfrac{S}{7}$

$\dfrac{}{7}$ $\dfrac{}{12}$ $\dfrac{}{4}$ $\dfrac{W}{13}$

$\dfrac{}{9}$ $\dfrac{}{1}$ $\dfrac{P}{3}$ $\dfrac{}{1}$ $\dfrac{}{8}$ $\dfrac{}{6}$ $\dfrac{}{1}$ $\dfrac{D}{11}$

$\dfrac{}{14}$ $\dfrac{}{8}$ $\dfrac{}{15}$ $\dfrac{}{6}$ $\dfrac{}{4}$ $\dfrac{}{9}$ $\dfrac{}{7}$.

Example: _____

NAME _____ **DATE** _____

 6.10

Use the numbers written below the spaces to discover the mystery message about a math idea. Each number represents a different letter of the alphabet. Some numbers are already shown with their respective letters. When you've decoded the message, write an example below the message to illustrate the math idea.

$$\frac{}{1} \quad \frac{P}{2} \quad \frac{}{3} \quad \frac{}{4} \quad \frac{}{5} \quad \frac{}{6} \quad \frac{}{7} \quad \frac{}{8}$$

$$\frac{}{4} \quad \frac{}{9} \quad \quad \frac{}{2} \quad \frac{}{3} \quad \frac{I}{10} \quad \frac{}{11} \quad \frac{}{12}$$

$$\frac{}{13} \quad \frac{}{6} \quad \frac{}{11} \quad \frac{}{14} \quad \frac{}{12} \quad \frac{}{3} \quad \frac{}{15} \quad \quad \frac{}{10} \quad \frac{}{15}$$

$$\frac{}{1} \quad \quad \frac{}{2} \quad \frac{}{3} \quad \frac{}{10} \quad \frac{}{11} \quad \frac{}{12}$$

$$\frac{}{9} \quad \frac{}{1} \quad \frac{}{7} \quad \frac{T}{8} \quad \frac{}{4} \quad \frac{}{3} \quad \frac{}{10} \quad \frac{Z}{16} \quad \frac{}{1}\text{-}$$

$$\frac{}{8} \quad \frac{}{10} \quad \frac{}{4} \quad \frac{}{13}.$$

Example: _____

6.11

Use the numbers written below the spaces to discover the mystery message about a math idea. Each number represents a different letter of the alphabet. Some numbers are already shown with their respective letters. When you've decoded the message, write an example below the message to illustrate the math idea.

$$\overline{\underset{1}{}}\ \overline{\underset{2}{}}\ \overline{\underset{3}{}}\qquad \overline{\underset{4}{}}\ \overline{\underset{5}{}}\ \overline{\underset{6}{O}}\ \overline{\underset{7}{}}\ \overline{\underset{8}{}}\ \overline{\underset{9}{}}\ \overline{\underset{1}{}}$$

$$\overline{\underset{6}{}}\ \overline{\underset{11}{}}\qquad \overline{\underset{1}{T}}\ \overline{\underset{12}{}}\ \overline{\underset{6}{}}$$

$$\overline{\underset{13}{N}}\ \overline{\underset{3}{}}\ \overline{\underset{14}{}}\ \overline{\underset{15}{}}\ \overline{\underset{1}{}}\ \overline{\underset{16}{}}\ \overline{\underset{17}{}}\ \overline{\underset{3}{}}$$

$$\overline{\underset{13}{}}\ \overline{\underset{8}{}}\ \overline{\underset{18}{}}\ \overline{\underset{19}{}}\ \overline{\underset{3}{E}}\ \overline{\underset{5}{}}\ \overline{\underset{20}{}}$$

$$\overline{\underset{3}{}}\ \overline{\underset{21}{}}\ \overline{\underset{8}{}}\ \overline{\underset{15}{}}\ \overline{\underset{22}{}}\ \overline{\underset{20}{}}\qquad \overline{\underset{15}{}}$$

$$\overline{\underset{4}{P}}\ \overline{\underset{6}{}}\ \overline{\underset{20}{}}\ \overline{\underset{16}{}}\ \overline{\underset{1}{}}\ \overline{\underset{16}{I}}\ \overline{\underset{17}{}}\ \overline{\underset{3}{}}$$

$$\overline{\underset{13}{}}\ \overline{\underset{8}{}}\ \overline{\underset{18}{}}\ \overline{\underset{19}{}}\ \overline{\underset{3}{}}\ \overline{\underset{5}{}}.$$

Example: _____

NAME _____ DATE _____

 6.12

Use the numbers written below the spaces to discover the mystery message about a math idea. Each number represents a different letter of the alphabet. Some numbers are already shown with their respective letters. When you've decoded the message, write an example below the message to illustrate the math idea.

$\dfrac{}{1}$ $\dfrac{}{2}$ $\dfrac{}{3}$ $\quad$ $\dfrac{}{4}$ $\dfrac{R}{5}$ $\dfrac{}{6}$ $\dfrac{}{7}$ $\dfrac{}{8}$ $\dfrac{}{9}$ $\dfrac{}{1}$

$\dfrac{}{6}$ $\dfrac{}{10}$ $\quad$ $\dfrac{}{11}$

$\dfrac{}{15}$ $\dfrac{}{3}$ $\dfrac{G}{17}$ $\dfrac{}{11}$ $\dfrac{}{1}$ $\dfrac{}{13}$ $\dfrac{}{14}$ $\dfrac{}{3}$

$\dfrac{}{11}$ $\dfrac{N}{15}$ $\dfrac{}{16}$ $\quad$ $\dfrac{}{11}$

$\dfrac{P}{4}$ $\dfrac{}{6}$ $\dfrac{}{12}$ $\dfrac{}{13}$ $\dfrac{}{1}$ $\dfrac{}{13}$ $\dfrac{}{14}$ $\dfrac{}{3}$

$\dfrac{}{15}$ $\dfrac{}{8}$ $\dfrac{}{18}$ $\dfrac{}{19}$ $\dfrac{}{3}$ $\dfrac{}{5}$ $\quad$ $\dfrac{}{13}$ $\dfrac{}{12}$

$\dfrac{}{15}$ $\dfrac{}{3}$ $\dfrac{}{17}$ $\dfrac{}{11}$ $\dfrac{}{1}$ $\dfrac{}{13}$ $\dfrac{}{14}$ $\dfrac{E}{3}$.

Example: _____

6.13

Use the numbers written below the spaces to discover the mystery message about a math idea. Each number represents a different letter of the alphabet. Some numbers are already shown with their respective letters. When you've decoded the message, write an example below the message to illustrate the math idea.

T __ __   __ __ __   __ __
1 2 3   4 5 6   7 8

__ W __   __ __ G __ __ __ __ __
1 15 7   9 3 10 11 1 12 13 3

__ __ __ __ __ __ __ S
12 9 1 3 10 3 14 4

__ __ __ __ __ __   __
3 16 5 11 17 4   11

__ __ __ __ __ __ __ __
9 3 10 11 1 12 13 3

__ __ __ __ __ __ R .
12 9 1 3 10 3 14

Example: _____

6.14

Use the numbers written below the spaces to discover the mystery message about a math idea. Each number represents a different letter of the alphabet. Some numbers are already shown with their respective letters. When you've decoded the message, write an example below the message to illustrate the math idea.

	Q								
12	13	14	15	7	3	12		16	6

					C					
9	16	8	12	3	9	14	1	5	17	3

							S
5	8	1	3	18	3	7	12

				E						
4	5	6	6	3	7		10	20		15	8

16	4	4		8	14	21	10	3	7

Example: _____

6.15

Use the numbers written below the spaces to discover the mystery message about a math idea. Each number represents a different letter of the alphabet. Some numbers are already shown with their respective letters. When you've decoded the message, write an example below the message to illustrate the math idea.

$$\frac{}{1} \quad \frac{}{9} \; \frac{E}{3} \; \frac{}{5} \qquad \frac{}{7} \; \frac{}{10}$$

$$\frac{O}{7} \; \frac{}{2} \; \frac{}{11} \; \frac{}{3} \; \frac{}{2} \; \frac{}{3} \; \frac{}{11}$$

$$\frac{P}{12} \; \frac{}{1} \; \frac{}{6} \; \frac{}{2} \; \frac{}{9} \qquad \frac{}{6} \; \frac{}{9}$$

$$\frac{}{13} \; \frac{}{1} \; \frac{L}{4} \; \frac{}{4} \; \frac{}{3} \; \frac{}{11} \qquad \frac{}{1}$$

$$\frac{R}{2} \; \frac{}{3} \; \frac{}{4} \; \frac{}{1} \; \frac{}{5} \; \frac{}{6} \; \frac{}{7} \; \frac{}{8}.$$

Example: _____

6.16

Use the numbers written below the spaces to discover the mystery message about a math idea. Each number represents a different letter of the alphabet. Some numbers are already shown with their respective letters. When you've decoded the message, write an example below the message to illustrate the math idea.

___ ___   ___ ___ ___ ___ ___ I ___ ___
1   2   3   4   2   5   6   1 7   2

___ ___ ___ ___ S ,   ___ Q ___ ___ ___
8   9   1   10   11   12   13 4   9   14

___ ___ ___ ___ ___   ___ E ___ ___ ___
3   1   10   11   6   6   12 10   15   11

___ ___ ___ ___   ___ ___ ___ ___
15   4   11   6   16   9   17   12

___ ___ ___ ___ ___   ___ ___ ___ ___ N ___
12   13   4   9   14   11   12   5   7   2 18

___ ___ ___ ___ ___ .
6   12   10   15   11

Example: _____

6.17

Use the numbers written below the spaces to discover the mystery message about a math idea. Each number represents a different letter of the alphabet. Some numbers are already shown with their respective letters. When you've decoded the message, write an example below the message to illustrate the math idea.

T __ __ __ L __ __ __ __ __
1 2   3 4 5 1 6 7 5 8

__ __ __ __ __ I __ __ __,
9 6 10 2 3 6 11 5 12

__ __ __   __ __ __
4 12 13   1 14 13

__ __ __ __ __ __ __ __ __ __ __ E
16 6 12 1 15 6 9 4 1 6 17 13

__ __ O __ __ __ __ __.
7 15 2 7 13 15 1 8

Example: _____

Copyright © 2010 by John Wiley & Sons, Inc., *The Algebra Teacher's Activity-a-Day*

6.18

Use the numbers written below the spaces to discover the mystery message about a math idea. Each number represents a different letter of the alphabet. Some numbers are already shown with their respective letters. When you've decoded the message, write an example below the message to illustrate the math idea.

__   __   __   __   __   __   __   __   __   __
2   1   8   2   3   6   2   11   10   4

E   __   __   __   __   __   __   __
7   1   8   2   11   10   12   9

__   __   __   __   __   C   __   __   __ -
10   5   2   5   7   4   12   9   3

D   __   __   __   __   __
3   7   15   6   7   7

__   Q   __   __   __   __   __   __ .
7   1   8   2   11   10   12   9

Example: _____

NAME _____ DATE _____

 6.19

Use the numbers written below the spaces to discover the mystery message about a math idea. Each number represents a different letter of the alphabet. Some numbers are already shown with their respective letters. When you've decoded the message, write an example or a response below the message to illustrate the math idea.

___ ___ ___   ___ L̲ ___ ___ ___   ___ ___
1   2   3   4 5 6   7   3   6   8

___   V̲ ___ ___ ___ ___ ___ ___ ___
9   10 3 11 1 12 13 9   5

___ ___ ___ ___   ___ ___ E̲ ___
5 12 14   3   15   6 3   4

___ ___ ___   ___ X̲ ___ ___ ___.
14 6   1   3 16 12 4   1

Example: _____

6.20

Use the numbers written below the spaces to discover the mystery message about a math idea. Each number represents a different letter of the alphabet. Some numbers are already shown with their respective letters. When you've decoded the message, write an example or a response below the message to illustrate the math idea.

| ___ | ___ | E | ___ | ___ | ___ | ___ | ___ |
| 8 | 6 | 5 | 10 | 7 | 1 | 11 | 5 |

| O | ___ | ___ | ___ | ___ | ___ | ___ |
| 1 | 2 | 9 | 7 | 12 | 4 | 5 |

| ___ | ___ | ___ | ___ | ___ | S | ___ | ___ | ___ | ___ |
| 5 | 15 | 14 | 9 | 7 | 10 | 3 | 12 | 10 | 5 |

| ___ | V | ___ | ___ | ___ | ___ | ___ |
| 1 | 13 | 5 | 3 | 3 | 14 | 4 |.

Example: _____

SECTION SEVEN

What Am I?

In the activities in this section, students must think about each clue provided in the problem. The first clue should generate a generalized set of ideas, to which the remaining clues will be applied. With each additional clue, students should be able to reduce the original set of ideas to a smaller set until finally a single, specific expression is obtained.

Example 7

Use the given clues to identify a number, expression, or equation. All three clues must be satisfied. Be specific.

➲ I am a monomial in the variable x.

➲ My degree is 3.

➲ My coefficient is a composite number less than 10 but divisible by 2 and 3.

What am I?

Explanation: First, students must identify a general monomial in x. This might be x, x^2, x^3, x^4, and so forth. This is the generalized set from which to begin reasoning. The second clue narrows the choices to x^3 in order to have a monomial of degree 3.

Any one of the original forms might have a coefficient besides $+1$; therefore, students must consider the third clue. They should first list all composite numbers less than 10: 4, 6, 8, and 9. These numbers are composite numbers because they have factors other than themselves and the number one. Students should then analyze each of these four numbers with respect to the possible factors 2 and 3. Only the number 6 has both factors; hence, 6 is the required coefficient. The final expression that satisfies all three clues of this particular activity is $6x^3$. Students have found their answer by applying logical reasoning.

Guide students to share their thought processes during a class discussion of the problem.

NAME _____ **DATE** _____

7.1

Use the given clues to identify a number, expression, or equation. All three clues must be satisfied. Be specific.

⮩ I am a cubic root of an even number.

⮩ My cube is less than 100.

⮩ I am a composite number.

What am I?

7.2

Use the given clues to identify a number, expression, or equation. All three clues must be satisfied. Be specific.

➲ I have only five factors in my prime factorization.

➲ I am between 60 and 75.

➲ The number 3 serves as exactly two of my prime factors.

What am I?

NAME _____   DATE _____

 7.3

Use the given clues to identify a number, expression, or equation. All three clues must be satisfied. Be specific.

➲ I name a point on the number line.

➲ My distance to the point, +2, is 5 units.

➲ I am a negative number.

What am I?

7.4

Use the given clues to identify a number, expression, or equation. All three clues must be satisfied. Be specific.

➲ I am a monomial in the variable x.

➲ My degree is 3.

➲ My coefficient is a composite number less than 10 but divisible by 2 and 3.

What am I?

NAME _____   **DATE** _____

7.5

Use the given clues to identify a number, expression, or equation. All three clues must be satisfied. Be specific.

➲ I am a monomial of first degree in x.

➲ I am relatively prime to 4ym.

➲ My coefficient is the mean value of 4, 3, and 8.

What am I?

7.6

Use the given clues to identify a number, expression, or equation. All three clues must be satisfied. Be specific.

➲ When I am added to 12x, our sum is 0.

➲ When I divide 12x, the quotient is –1.

➲ When I am added to 13x, our sum is x.

What am I?

7.7

Use the given clues to identify a number, expression, or equation. All three clues must be satisfied. Be specific.

➲ I am the fourth power of a base.

➲ My base has the variable factors x and y.

➲ The two digits of my coefficient have a sum of 9 when I am simplified.

What am I when simplified?

7.8

Use the given clues to identify a number, expression, or equation. All three clues must be satisfied. Be specific.

➲ I make −4(x + 2) equal −4x − 8.

➲ I make $3x^2 + 6x$ equal 3x(x + 2).

➲ I make 25(18) = 25(10) + 25(8).

What am I?

7.9

Use the given clues to identify a number, expression, or equation. All three clues must be satisfied. Be specific.

➲ I am a second-degree trinomial in x.

➲ One of my prime factors is (x + 2).

➲ The trinomial $x^2 + 6x + 9$ is divisible by my other prime factor.

What am I?

7.10

Use the given clues to identify a number, expression, or equation. All three clues must be satisfied. Be specific.

➲ I am a quadratic equation in x.

➲ My roots are −5 and +3.

➲ The coefficient of my quadratic term is 3.

What am I?

7.11

Use the given clues to identify a number, expression, or equation. All three clues must be satisfied. Be specific.

➲ I am a binomial of the form x + c, where c is a constant.

➲ My absolute value equals 5.

➲ The solutions of my absolute value equation are +7 and −3.

What am I?

7.12

Use the given clues to identify a number, expression, or equation. All three clues must be satisfied. Be specific.

➲ I am a linear function f of x for all real x.

➲ I contain the ordered pairs (+3, −1) and (−3, +1).

➲ My slope is negative.

What function am I? (Give rule for f.)

NAME _____ **DATE** _____

 7.13

Use the given clues to identify a number, expression, or equation. All three clues must be satisfied. Be specific.

➲ I am a function of x where f(x) ≥ 0.

➲ My graph is a parabola with the y-axis or vertical axis as its line of symmetry.

➲ My vertex is at the origin and I contain the ordered pair (−2, +12).

What function am I? (Give rule for f.)

NAME ______________________ DATE ______________________

7.14

Use the given clues to identify a number, expression, or equation. All three clues must be satisfied. Be specific.

- My graph is a line.
- All the independent values (abscissas) of my coordinate pairs have +4 as the dependent value (ordinate).
- My slope is zero.

What function am I? (Give rule for f.)

NAME _____ **DATE** _____

 7.15

Use the given clues to identify a number, expression, or equation. All three clues must be satisfied. Be specific.

➲ My graph is a line and I am a relation but not a function.

➲ As my dependent values increase, the independent or horizontal value does not change.

➲ One of my ordered pairs is (−2, −4).

What am I? (Give rule.)

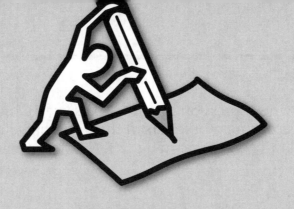

SECTION EIGHT

Al-ge-grams

In the activities in this section, students must apply the order of operations and other mathematical procedures accurately to simplify algebraic expressions. Once simplified, the remaining expression will reveal a special message. The message is general and not necessarily mathematical. As students move through the required processes, they should maintain the left-right order of their numerator letters. This will reduce the need to rearrange a few letters, and some numbers as well, in order to find the hidden message.

Example 8

Simplify the following expression in order to discover a hidden message. When possible, keep numerator letters in their original left-right order. It may be necessary at the end to rearrange a few letters or numbers to find the message.

Hint: Believe it or not!

$$\frac{am}{3}\left(\frac{th}{5} \div \frac{1}{5}\right)\left(15sa \div \frac{3a}{i}\right)\left(\frac{3tnu}{30} \cdot 6ofo\right) = ?$$

Explanation: To simplify this expression, students must first perform the operation within each set of parentheses. A string of fractional factors will result, which can be reduced by dividing or removing common factors in the numerators and denominators. At this point, students should try to preserve the initial order of all the letters in the numerators; letters or variables should not be commuted. This will help with identifying the message later.

$$\frac{am}{3}\left(\frac{th}{5} \cdot \frac{5}{1}\right)\left(15sa \cdot \frac{i}{3a}\right)\left(\frac{3tnu}{30} \cdot 6ofo\right) =$$

$$\frac{am}{3}\left(\frac{th}{1}\right)(5s \cdot i)\left(\frac{tnu}{10} \cdot 6ofo\right) =$$

$$\frac{am}{3}(th)(si)\left(\frac{tnu}{2} \cdot 6ofo\right) =$$

$$am(th)(si)(tnuofo)$$

When the reduction process is complete, all that should remain are the letter groupings **am(th)(si)(tnuofo)**, which after unscrambling will translate into the final message, "math is fun too."

Some problems in this section will involve operations with radicals and powers, as well as the factoring of various trinomials. In the final step, any power should be expanded into its separate factors, for example, $b^3 = bbb$, to aid the unscrambling process. Students must be extremely accurate with algebraic manipulations in order for their hidden messages to be easily recognized.

8.1

Simplify the following expression in order to discover a hidden message. When possible, keep numerator letters in their original left-right order. It may be necessary at the end to rearrange a few letters or numbers to find the message.

Hint: Believe it or not!

$$\frac{am}{3}\left(\frac{th}{5} \div \frac{1}{5}\right)\left(15sa \div \frac{3a}{i}\right)\left(\frac{3tnu}{30} \cdot 6ofo\right) = ?$$

8.2

Simplify the following expression in order to discover a hidden message. When possible, keep numerator letters in their original left-right order. It may be necessary at the end to rearrange a few letters or numbers to find the message.

Hint: A quality you possess.

$$\left(\frac{7A}{5M} \div \frac{13}{5}\right)\left(\frac{39U}{18R} \cdot \frac{T}{9A}\right)\left(\frac{54R^2M6Y}{7} \cdot \frac{BZ}{3BT}\right) = ?$$

8.3

Simplify the following expression in order to discover a hidden message. When possible, keep numerator letters in their original left-right order. It may be necessary at the end to rearrange a few letters or numbers to find the message.

Hint: A pleasant reception.

$$\frac{D^2}{30K}\left[\frac{40WEQ}{4} \cdot \frac{LK}{5Q}\right]\left[\frac{5CGOA}{3G} \div \frac{A}{3ME}\right]\left[\frac{6AC}{D} \cdot \frac{2KB}{4D}\right] = ?$$

8.4

Simplify the following expression in order to discover a hidden message.
When possible, keep numerator letters in their original left-right order.
It may be necessary at the end to rearrange a few letters or numbers to find
the message.

Hint: A great attitude to have.

$$\left[\frac{2I}{10} \div \frac{1}{5UD} \right] \div \frac{6D}{2ML} \left[\frac{20VT}{6} - \frac{2VT}{6} \right] AH = ?$$

8.5

Simplify the following expression in order to discover a hidden message.
When possible, keep numerator letters in their original left-right order.
It may be necessary at the end to rearrange a few letters or numbers to find
the message.

Hint: A good answer to have.

$$\left(\frac{4R}{U} \div \frac{W}{(JU)^2}\right)\left(\frac{ST}{2J} - \frac{ST}{3J}\right)\left(\frac{SA^2}{Y} \cdot \frac{Y^2N}{A}\right)\left(\frac{WO}{2R} + \frac{WO}{R}\right) = ?$$

8.6

Simplify the following expression in order to discover a hidden message. When possible, keep numerator letters in their original left-right order. It may be necessary at the end to rearrange a few letters or numbers to find the message.

Hint: Brighten my day!

$$\frac{2SWX}{B}\left(\frac{MIBX}{4} - \frac{MIBX}{8}\right)\left(\frac{LW}{EV} \div \frac{(WX)^2}{9}\right)\left(\frac{EVE}{3} + \frac{EVE}{9}\right) = ?$$

8.7

Simplify the following expression in order to discover a hidden message. When possible, keep numerator letters in their original left-right order. It may be necessary at the end to rearrange a few letters or numbers to find the message.

Hint: OK is the answer to this.

$$\frac{h}{3x}\left(2xo - \frac{xo}{2}\right)\left(wz \div \frac{2z}{4r}\right) + uv^0 = ?$$

8.8

Simplify the following expression in order to discover a hidden message.
When possible, keep numerator letters in their original left-right order.
It may be necessary at the end to rearrange a few letters or numbers to find
the message.

Hint: Gently, please!

$$\left[2(6th)x \div \frac{2(3x)}{mgu} \right] \left[\frac{e}{t} - \frac{2e}{4t} \right] = ?$$

NAME _____ DATE _____

 **8.9**

Simplify the following expression in order to discover a hidden message. When possible, keep numerator letters in their original left-right order. It may be necessary at the end to rearrange a few letters or numbers to find the message.

Hint: When you study every day, _____.

$$\frac{5E}{11}\left[\frac{4A}{5}+\frac{2A}{25}\right]\left(\frac{5IH}{2}\right)\left[\frac{3TS}{4}+\frac{2TS}{8}\right]\left(\frac{3Z}{4}-\frac{2Z}{3}\right)(12M)=?$$

8.10

Simplify the following expression in order to discover a hidden message.
When possible, keep numerator letters in their original left-right order.
It may be necessary at the end to rearrange a few letters or numbers to find
the message.

Hint: A good thing to know.

$$\frac{4AP}{5}\left[\frac{3PL}{4} \div \frac{XP}{5GE}\right]\left[\frac{(\sqrt{RB})^2}{3} - \frac{\sqrt{(RB)^2}}{6}\right]\left(2AX \cdot \frac{I}{P}\right) = ?$$

Al-ge-grams 8.10

8.11

Simplify the following expression in order to discover a hidden message. When possible, keep numerator letters in their original left-right order. It may be necessary at the end to rearrange a few letters or numbers to find the message.

Hint: Who wins the prize?

$$\frac{r}{3}\left[\frac{(\sqrt{ve})^2}{4} + \frac{3(\sqrt{ve})^2}{4}\right](6ey)\left[\left(\frac{o^2e}{c}\right) \div \left(\frac{2u^3}{n^2}\right)\right](u^4c^2ts) = \,?$$

8.12

Simplify the following expression in order to discover a hidden message. When possible, keep numerator letters in their original left-right order. It may be necessary at the end to rearrange a few letters or numbers to find the message.

Hint: This will take you far in life.

$$\sqrt{\frac{4M^2X^2}{9T^2}}\left[\frac{(5RE)^2}{30}\right]\left[\frac{3T}{2}+\frac{3T}{4}\right]\left[\frac{TO}{3X}\div\frac{RE}{2HA}\right]\left(\frac{2PW+4PW}{5}\right)=?$$

NAME _____   DATE _____

8.13

Simplify the following expression in order to discover a hidden message. When possible, keep numerator letters in their original left-right order. It may be necessary at the end to rearrange a few letters or numbers to find the message.

Hint: The results of good study habits.

$$\left[\frac{3w^2}{4vk} \div \frac{3wn}{i}\right]\left[(u+v)^2 - (u-v)^2\right]\left(\frac{(kn)^2}{k}\right) = ?$$

8.14

Simplify the following expression in order to discover a hidden message. When possible, keep numerator letters in their original left-right order. It may be necessary at the end to rearrange a few letters or numbers to find the message.

Hint: What you are!

$$\frac{4abrt}{b^2 - 2b}\left(\frac{b}{2rt} - \frac{1}{rt}\right)\left[5wenh \div \frac{10nh}{s}\right]\left(\frac{od}{6} + \frac{2o}{3}\right)\left(\frac{18me}{3d + 12}\right) = ?$$

NAME _____ DATE _____

8.15

Simplify the following expression in order to discover a hidden message. When possible, keep numerator letters in their original left-right order. It may be necessary at the end to rearrange a few letters or numbers to find the message.

Hint: A special someone.

$$\left(\frac{3U}{2X} \div \frac{4E}{R}\right)\left(\frac{4(\sqrt[3]{27})}{X} + \frac{5\sqrt{16}}{X}\right)\left(\frac{XM}{E} \cdot \frac{XE^2}{3}\right)$$

$$\cdot \left[\left(\frac{2X^2 + 3X - 9}{X + 3}\right)\left(\frac{E}{2X - 3}\right)\right] = ?$$

8.16

Simplify the following expression in order to discover a hidden message. When possible, keep numerator letters in their original left-right order. It may be necessary at the end to rearrange a few letters or numbers to find the message.

Hint: A special alert!

$$\left[\frac{W(X^2 - 5X + 6)}{BP(X + 5)} \div \frac{(X - 3)}{A(X^2 + 8X + 15)LK} \right] (8 \text{ DOWN})$$

$$\cdot \left(\frac{T}{8W} \right) \cdot \frac{RBP}{X^2 + X - 6} \left[\frac{(FUNNY)^2}{UN} \div (FNY)^2 \right] = ?$$

8.17

Simplify the following expression in order to discover a hidden message. When possible, keep numerator letters in their original left-right order. It may be necessary at the end to rearrange a few letters or numbers to find the message.

Hint: Why keep trying?

$$3U^2R^2\left(\frac{C}{6} \div \frac{UR}{2AP}\right)\left[\frac{(A+3)(A-3)+9}{A}\right] \cdot \left[\left(\frac{B^2-X^2}{B+X}\right)+X\right]\left(\frac{\sqrt{L^4E^4}}{LE}\right) = \,?$$

8.18

Simplify the following expression in order to discover a hidden message. When possible, keep numerator letters in their original left-right order. It may be necessary at the end to rearrange a few letters or numbers to find the message.

Hint: What you can say at the end of the school day.

$$\left(\frac{12d^2e}{11td}\right)\left(\frac{du+3u}{d^2+5d+6}\right)\left[(d^2-4)\div\frac{(d-2)}{am}\right]\left(\frac{2t^2i}{3}+\frac{t^2i}{4}\right)=?$$

8.19

Simplify the following expression in order to discover a hidden message. When possible, keep numerator letters in their original left-right order. It may be necessary at the end to rearrange a few letters or numbers to find the message.

Hint: You have high expectations for this.

$$\left(\frac{a^3p^4h}{a^2p^2}\right) \div \left(\frac{(y-8)^2}{y^3 - 16y^2 + 64y}\right) \left[\frac{p^2vxy}{x^2} \div \frac{p^2y}{ax}\right] (-k) \left(\frac{-t^2i^2m}{itm}\right) \div \left(\frac{no^{-2}}{n^2o^{-1}}\right) = ?$$

8.20

Simplify the following expression in order to discover a hidden message. When possible, keep numerator letters in their original left-right order. It may be necessary at the end to rearrange a few letters or numbers to find the message.

Hint: A good ending!

$$\left(\frac{\sqrt[3]{a^4h^4}}{ve} \cdot \frac{1}{(ah)^{1/3}} \right) \div (ev)^{-2} n^0 a \left(\frac{(n+1)!}{n!} - 1 \right)$$

$$\cdot\, i^3 e \sqrt{m} \left(\frac{1}{c} \right)^{-1} i^{-2} \left(\frac{1}{\sqrt{m}} \right) usm^2 \left(\frac{e\,(r+1)!}{r+1} \right) = ?$$

SECTION NINE

Potpourri

In the activities in this section, students will be involved with three different types of problems: cooperative games, oral team problems, and mini-investigations. This section of problems encourages students to work together and to communicate their ideas to one another. Students gain experience with a variety of strategies needed in problem solving.

In *cooperative games* (activities 9.1 to 9.5), students work with one other student to solve special problems. Tactile experiences are provided through the use of manipulatives. Scissors will be needed to cut out game pieces for several of the games.

In solving *oral team problems* (activities 9.6 to 9.10), students will work in teams of two to four students to solve a problem. They must have the skills needed to solve the particular problem assigned. The students must work without calculators or pencil and paper. The solving process must be oral, although hand movements are allowed. The emphasis is on mental mathematics and any numerical shortcuts that students may think to use. Four possible responses are provided for each problem; three may be eliminated through logical reasoning.

In solving *mini-investigations* (activities 9.11 to 9.15), students may work independently or with other students to explore or test various mathematical statements. For some problems, students must decide when a statement is false or true. They must be able to provide examples or counterexamples. Number patterns and other strategies, such as making tables or creating easier problems, will be useful in solving many problems. Occasionally, simple algebraic concepts will be applied in a geometric setting.

At the beginning of each of the three categories of this section, a sample problem is presented. If time permits, after a Potpourri activity has been completed, students should be encouraged to share their answers and their reasoning and strategies with the entire class.

Example for Activity 9.3

Gameboard:

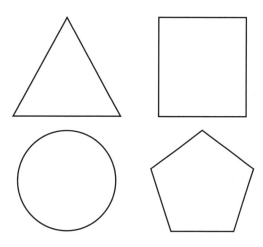

Explanation: Two students will cut out the four cards on the second page of the activity and share the cards equally. As they take turns describing the equation on each card, the students will place small counters (such as unit cubes or small buttons) inside the shapes on the gameboard to represent the values of the shapes when the amounts satisfy the equation on the card. For example, one card contains "circle + square + triangle = 17." Students might place 4 counters inside the circle of the gameboard, 5 counters inside the square, and 8 counters inside the triangle. These three amounts total 17 counters, as required by the equation on the card. However, these amounts will not satisfy the equation on another card, "square − circle = triangle − 5," because $5 - 4 \neq 8 - 5$. So students must rearrange the initial amounts of counters selected for the first card until they satisfy the equation of the second card as well as that of the first card. This process continues as each new card's equation is considered.

When *all* equations are finally satisfied simultaneously, the counters lying inside the four shapes represent the solution. All four equations on the different cards will be satisfied when the following amounts occur: 3 counters are in the circle, 4 are in the pentagon, 8 are in the triangle, and 6 are in the square. These results should be recorded in the blank on the activity sheet. Students have basically solved a system of equations. The experience of moving the counters from one shape to another helps students to understand how a set of different equations can have a common solution.

9.1

Mystery Matrix: Work with a partner. Cut out sixteen small paper squares and label each square with its own number, using 1 through 16. Arrange the labeled squares in the boxes of the following grid so that the sum of each of the four columns is the same. Can you find more than one possible arrangement for the paper squares? Record each grid arrangement of numbers that you find.

9.2

Work with a partner.

1. With no two fractions being equal and using at least seven fractions, make an addition problem whose sum is 1. Solve this problem by using a physical model; that is, fold a long strip of adding machine tape (approximately 12 inches long) into smaller fractional segments or parts and label the parts with their appropriate fraction names. All folds will be parallel to each other. Assume that the total length of the paper strip equals 1 in value. When finished, share your folding and labeling strategies with others in the class.

Record your final sum of fractions here: 1 = _____

Draw a diagram of your folded and labeled strip of paper here:

2. *Extension:* At a meat counter, people must take numbers to be served. (a) Give to each of 20 people a different rational number between −1.0 and +0.1. Arrange these numbers to show the order in which the people will be served. (b) Can you find another set of 20 numbers different from the first set? (*Hint:* Try to use a mixture of both decimal and fractional numbers for this problem.)

9.3

Cut out the four cards on the following page and share them equally with one other student. As you take turns describing the equation on each card, place small counters (such as unit cubes or small buttons) inside the shapes on the gameboard to represent the values of the shapes when the amounts satisfy the equation of shapes being described. Move counters as needed for the various equations being considered until the amounts in all shapes finally satisfy *all* equations simultaneously.

Record the final shape values found: _____

Gameboard:

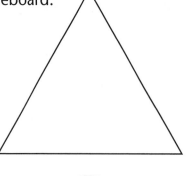

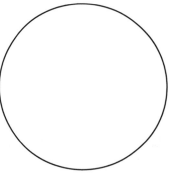
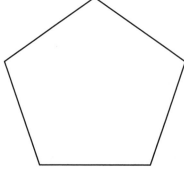

Potpourri 9.3 (continued)

Cut these four cards apart and distribute them equally.

$\bigcirc + \square + \triangle = 17$

Find each shape's value.

$\bigcirc + \pentagon + \triangle = 15$

Find each shape's value.

$\square - \bigcirc = \triangle - 5$

Find each shape's value.

$\bigcirc \times \triangle = \square \times 4$

Find each shape's value.

9.4

Work with a partner.

Situation: Bill Durr was a contractor who built patios and pools. He had been hired to make a square patio using tiles of rectangular slate that were lying scattered in the backyard. After several hours of trying to arrange the heavy slate to form a square, Bill Durr quit.

1. Can you use the tiles of rectangular "slate" provided on the next page to solve the problem of the square patio? Cut out the tiles and try to arrange them into a square.
Draw a diagram here to show your result:

2. Assign letters a, b, and c to the three lengths involved in the "slate" tiles and label all lengths accordingly on your diagram. Can you tell what mathematical relationship is represented by your "slate" solution or diagram? (*Hint:* Think about the different areas involved and the total area of your diagram; represent each area algebraically.)
Record the equation for the relationship here:

Potpourri 9.4 (continued)

Patterns for "slate" tiles (cut out and use):

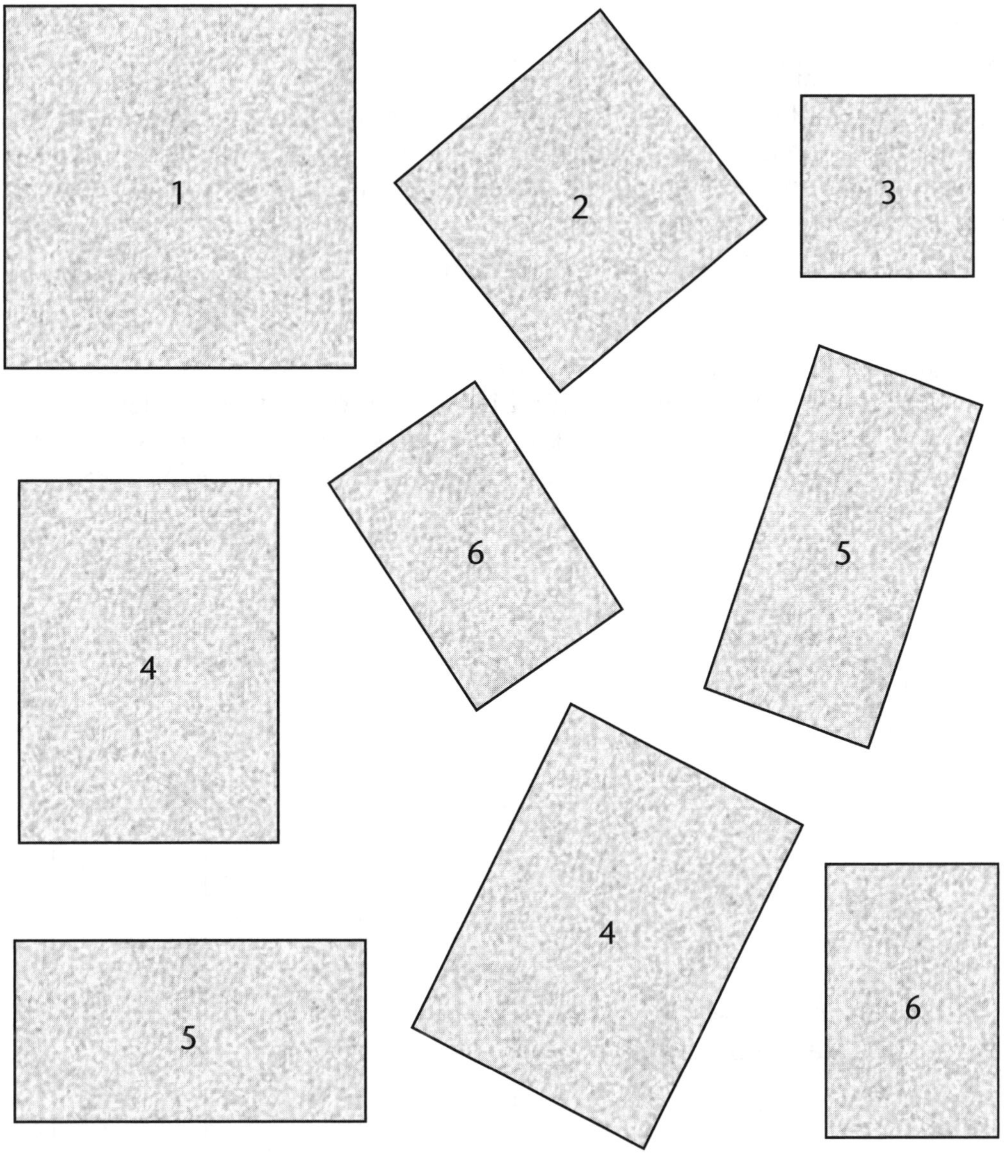

1

2

3

6

5

4

4

5

6

NAME _____ DATE _____

 9.5

Work with a partner.

Situation: Mrs. Jordan was trying to have her class work an algebra problem using manipulatives. She had four rectangular tiles to use in her demonstration. However, because she had not understood the problem herself, she was having a difficult time remembering how to show the students what to do. Can you help her?

Problem: Show $(a + b)^2 - (a - b)^2 = 4ab$, using the four tiles having dimensions a and b.

  1. Cut out the four tiles shown below.
  2. Using the four tiles, build a visual interpretation of the problem. Draw a diagram to record your tile structure. Be ready to explain your reasoning to the entire class. (*Hint:* Think of areas for the squared expressions as well as for the tiles themselves.)
    Diagram:

Cut out these
four tiles to use.

b	b	b	b
a	a	a	a

Example for Activity 9.7

In the following figure, m∠AEC = 70 degrees, m∠BED = 80 degrees, and m∠AED = 110 degrees. Find m∠BEC.

a. 10   b. 20   c. 30   d. 40

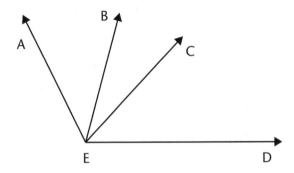

Explanation: Two to four students will work together to solve this problem. They must work without calculators or pencil and paper. The team must discuss the solving process orally, although hand movements are allowed. They must select the best of four given responses.

To solve the problem, students must notice that the measure of angle BEC is added twice when the measures of angles AEC and BED are added in 70 + 80 = 150, but the total measure of angle AED is 110, which indicates that the extra 40 comes from the extra measure of angle BEC. The answer is (d).

9.6

Work with a team of two to four students. The problem must be solved only mentally and orally. Do not use calculators or pencil and paper. Hand movements among team members are allowed. Select the best of four responses provided.

In the exercise 397 − 159, if 397 is increased by 10 and 159 is decreased by 5, in what way has the original difference been changed?

a. increased by 5   b. increased by 15

c. decreased by 5   d. no change

9.7

Work with a team of two to four students. The problem must be solved only mentally and orally. Do not use calculators or pencil and paper. Hand movements among team members are allowed. Select the best of four responses provided.

In the following figure, m∠AEC = 70 degrees, m∠BED = 80 degrees, and m∠AED = 110 degrees. Find m∠BEC.

a. 10   b. 20   c. 30   d. 40

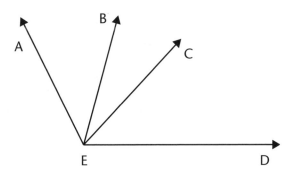

9.8

Work with a team of two to four students. The problem must be solved only mentally and orally. Do not use calculators or pencil and paper. Hand movements among team members are allowed. Select the best of four responses provided.

Given the following two circles, the small circle's area is approximately what fractional part of the large circle's area?

a. 1/9   b. 1/4   c. 1/3   d. 1/2

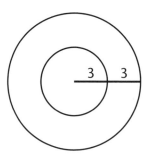

NAME _____ DATE _____

 9.9

Work with a team of two to four students. The problem must be solved only mentally and orally. Do not use calculators or pencil and paper. Hand movements among team members are allowed. Select the best of four responses provided.

Solve for x and y in the following equation, where the letter *i* represents the positive unit in complex numbers and x and y are real numbers:

$$3x + 1 + xy\,i = 13 + 20\,i$$

a. $x = 2, y = 7$ b. $x = 4, y = 5$ c. $x = 4, y = 2$ d. $x = 6, y = 5$

9.10

Work with a team of two to four students. The problem must be solved only mentally and orally. Do not use calculators or pencil and paper. Hand movements among team members are allowed. Select the best of four responses provided.

Graph $y = (1/3) x + 2$ mentally. Then choose the best of the following graphs for representing the given function.

a.

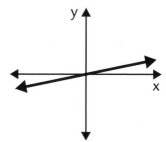

b.

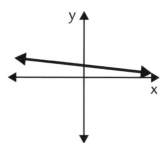

c.

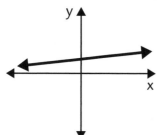

d.

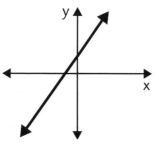

Example for Activity 9.12

Work independently or with other students.

1. Find the product if all letters of the alphabet are used and all are real numbers.

$$(k - a)(k - b)(k - c)(k - d) \ldots (k - y)(k - z) = ?$$

Because all letters of the alphabet are used, the factor $(k - k)$ exists and equals 0. When a factor of 0 occurs, the entire product will equal 0.

2. Find the given product.

$$\left(1 - \frac{1}{2}\right)\left(1 - \frac{1}{3}\right)\left(1 - \frac{1}{4}\right) \cdots \left(1 - \frac{1}{98}\right)\left(1 - \frac{1}{99}\right)\left(1 - \frac{1}{100}\right) = ?$$

Simplify each parenthesis group, then divide or remove common factors from the numerators and denominators. Here is the final product:

$$\frac{1}{2} \cdot \frac{2}{3} \cdot \frac{3}{4} \cdot \ldots \cdot \frac{98}{99} \cdot \frac{99}{100} = \frac{1}{100}$$

3. How many segments can be drawn between each pair of ten points? No three of the points are collinear or in a line together.

Explanation: Start with 2 points connected by 1 segment. Add another point not collinear to the first two and connect it to the other two points; there will be 3 segments in all. Add a fourth point not collinear to the other three and connect it to each of the other three points. Three more segments are drawn that do not coincide with the previous 3 segments, making 6 segments total. A fifth point will connect to the previous four points with 4 new segments, producing 10 segments total. Continue this process, if you wish, up to ten points. After drawing the first 3 or 4 points and their segments, however, you might make a table comparing number of points to total segments drawn for that number of points and extend the pattern shown in the segments column to find a total of 45 segments drawn for ten points.

9.11

Work independently or with other students.

1. If k − 8 is a negative real number and k is an integer, what is the largest possible value of k?

2. If k − 5 = m + 7, which is greater: k or m?

3. In 3 × 126, if 126 is increased by 25 and the factor 3 is doubled, how will the original product change? (*Hint:* Consider the new product, (2 × 3) (126 + 25), and apply the distributive property.)

NAME _____ **DATE** _____

 9.12

Work independently or with other students.

1. Find the product if all letters of the alphabet are used and all are real numbers.

$$(k - a)(k - b)(k - c)(k - d) \ldots (k - y)(k - z) = ?$$

2. Find the given product.

$$\left(1 - \frac{1}{2}\right)\left(1 - \frac{1}{3}\right)\left(1 - \frac{1}{4}\right) \cdots \left(1 - \frac{1}{98}\right)\left(1 - \frac{1}{99}\right)\left(1 - \frac{1}{100}\right) = ?$$

3. How many segments can be drawn between each pair of ten points? No three of the points are collinear or in a line together. (*Hint:* Make a table comparing various numbers of points used to the number of segments drawn using those points; in other words, make easier problems.)

9.13

Work independently or with other students.

Decide if each of the following statements is *always true*, *never true*, or *sometimes true*. If *always true*, give an example; if *never true*, give a counterexample. If *sometimes true*, give specific examples of when it is true and when it is false.

1. $x^{2k} \geq 0$ for all real x and integer k.

Circle one: *always/never/sometimes*
Examples or counterexamples:

2. The line for $y = ax + b$ has an x-intercept for x, y, a, and b as real numbers.

Circle one: *always/never/sometimes*
Examples or counterexamples:

3. If (a, c) is a solution to a linear equation $y = mx + b$, then (c, a) is also a solution.

Circle one: *always/never/sometimes*
Examples or counterexamples:

NAME _____ **DATE** _____

 9.14

Work independently or with other students.

Decide if each of the following statements is *always true, never true,* or *sometimes true*. If *always true*, give an example; if *never true*, give a counterexample. If *sometimes true*, give specific examples of when it is true and when it is false.

1. x^n for any real $x < 0$ and positive integer n will be negative.

Circle one: *always/never/sometimes*
Examples or counterexamples:

2. $(x + y)^2 = x^2 + y^2$ for all real numbers x and y.

Circle one: *always/never/sometimes*
Examples or counterexamples:

3. $x^2 = |-x|^2$ for every real x.

Circle one: *always/never/sometimes*
Examples or counterexamples:

NAME _____ DATE _____

 9.15

Work independently or with other students.

1. For the circular shaded region, give or estimate the coordinates of a point for each of the following locations:

a. outside the region

b. inside the region

c. on the boundary of the region (other than the labeled points)

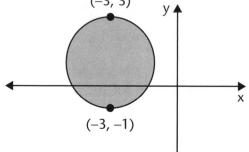

2. A square is inscribed in a right triangle with base b and height h so that one side of the square is parallel to the base of the triangle. Find the length of a side of the square in terms of b and h. (*Hint:* Consider the various areas involved.)

3. Consider a non-right triangle with all other conditions from problem 2. Find the square's side length in terms of b and h.

SECTION TEN

Calculator Explorations

In the activities in this section, students will be involved with two different types of problems: applications and graphical explorations. This section encourages students to work together and to communicate their ideas to one another. Students gain experience with data generation and analysis, and with interesting applications that can be simulated in the classroom.

In solving *applications* (activities 10.1 to 10.5), students work independently or with a partner to solve special problems. The regular calculator will be used to generate data from which patterns may be identified or conclusions drawn.

In solving *graphical explorations* (activities 10.6 to 10.15), students will work with a partner to complete each activity on a graphing calculator. Some activities will require finding a predictor equation for a set of data. Others will explore changes in functions and the effects of those changes on the graphs of the functions. Some of the functions may be new to students, but they still can explore changes in an unfamiliar "parent" form by means of the graphing calculator.

At the beginning of each of the two categories of this section, a sample activity is presented and discussed. If time permits, after an activity has been completed, students should be encouraged to share their answers and their reasoning and strategies with the entire class.

Example for Activity 10.3

Explanation: In this activity, students will use diagrams and symbolic notation to help them understand algebraic procedures. They will start by choosing any whole number, say 27. This number is represented by a small non-square rectangle and by X. The next steps will be shown as follows:

Add 3.
(30)

$X + 3$

Multiply by 2.
(60)

$2(X + 3) = 2X + 6$

Add 8.
(68)

$(2X + 6) + 8 = 2X + 14$

Divide by 2.
(34)

$(2X + 14) \div 2 =$
$(1/2)(2X + 14) =$
$X + 7$

Subtract start number.
(7)

$(X + 7) - X = 7$

Students should discover that the start number has now been removed, so only 7 is left. Everyone will have 7 left, no matter which number starts the procedure.

10.1

Work independently or with a partner. In each of the following exercises, look for a pattern among the numbers represented by the three given equations. Predict the next two equations and record them in the blanks. Use a calculator to confirm your results. Be ready to share the patterns you find with the entire class.

1. $1 \times 8 + 1 = 9$
    $12 \times 8 + 2 = 98$
    $123 \times 8 + 3 = 987$

    _____

    _____

2. $1 \times 9 + 2 = 11$
    $12 \times 9 + 3 = 111$
    $123 \times 9 + 4 = 1111$

    _____

    _____

3. $9 \times 9 + 7 = 88$
    $98 \times 9 + 6 = 888$
    $987 \times 9 + 5 = 8888$

    _____

    _____

10.2

Work independently or with a partner. In each of the following exercises, look for a pattern among the numbers represented by the three given equations. Predict the next two equations and record them in the blanks. Use a calculator to confirm your results. Be ready to share the patterns you find with the entire class.

1. $1 \times 1 = 1$
    $11 \times 11 = 121$
    $111 \times 111 = 12321$

    _____

    _____

2. $15 \times 15 = 225$
    $25 \times 25 = 625$
    $35 \times 35 = 1225$

    _____

    _____

3. $1(2^0 + 2^1 + 2^2) = 2^3 - 1 = 7$
    $2(3^0 + 3^1 + 3^2) = 3^3 - 1 = 26$
    $3(4^0 + 4^1 + 4^2) = 4^3 - 1 = 63$

    _____

    _____

10.3

Work with a partner. Use a calculator if necessary.

Complete column 2 and column 3 in the following table according to the information given in column 1. In column 2 use a small non-square rectangle as the tile for the unknown start number; use a small square as the tile for the value + 1 or 1.

Procedural Steps to Follow	Draw Tiles to Show Each Step	Write the Symbols to Show Each Step; Simplify
Start by choosing a number.		Let X represent the start number chosen. X
Add 3.		X + 3
Multiply by 2. (Draw the new tile set twice.)		2(X + 3) =
Add 8.		
Divide by 2. (Separate total tiles into 2 equal groups; keep 1 group.)		(2X + 14) ÷ 2 =
Subtract the start number.		
What is the result?		

Can you create another procedure like the one given here, then draw the tiles and write the symbols for each step?

10.4

Work with a partner to complete the following table. Use a calculator if necessary. Typically an input value follows the sequence of operations listed from *left* to *right* in the table. Use trial and error on exercise 1 to select and test a number from the given input set to see if the given output number is reasonable. Then reverse the sequence of operations (*right* to *left*) and apply their inverse operations to the output value to find the corresponding input value. Do not use parentheses on the calculator. For programmed commands SQUARE and SQUARE ROOT, apply the EQUALS key before selecting the command.

	Started with	Input	Operations	Output	Correct input exists?
1	whole number 0 to 15		+5; ×3; −6; ×2	36	
2	whole number 0 to 10		×2; +5; −8; +10	20	
3	negative rational −15 to −1		+4; −4; ×6	−36	
4	rational number 0 to 15		+5; ×0; +6; ÷2	3	
5	integer 0 to 15		×2; ×20; ×0.5; ÷2	5	
6	rational number −15 to 15		×2; +40; ×0.75; +5	20	
7	whole number		square; +3; sq root; −2	0	
8	whole number 0 to 10		+8; ×6; −4; ÷4	8	
9	integer −15 to −1		+10; ÷2; ×0; +1	1	
10	whole number		sq root; ×10; +8	9	

10.5

Work with a partner. Use a calculator to generate any data needed.

Situation: Your parents want to buy you a new 20-inch standard color TV for your room, but they also want to test your math skills. They want you to figure out the screen's dimensions for maximum area yet maintain the screen's 20-inch diagonal. It is time for a calculator!

Complete the following table for screen side lengths S1 (integers) and S2 (decimals) using a diagonal length of 20 inches for each pair of sides. Draw a diagram of the screen and label the sides and the diagonal. What equation will you need to find S2 in terms of S1 and 20? Record it at the top of the S2 column.

S1	S2 = _____	Area = S1 × S2
1		
2		
3		
4		
5		
6		
7		
8		
9		
10		
11		
12		
13		
14		
15		

Circle the row of the table that contains the maximum area computed.

Example for Activity 10.6

Situation: A ceramic mug containing tea was heated in a microwave oven, then placed on the kitchen counter and allowed to cool. Its temperature was measured every minute for six minutes and the following data were obtained:

Minutes Out of Oven	Temperature (C°)
0	87
1	83
2	80
3	78
4	75
5	72
6	70

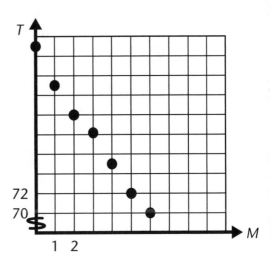

Explanation: Students will work with a partner to plot the data pairs on the grid provided and to plot the data pairs on a graphing calculator grid. In exercise 3, students will be asked to describe an equation that best fits their data. Answers and reasons will vary, but they should notice that the scatter plot appears to be linear. For example, the data appear to have a negative slope. Using (3, 78) and (4, 75) to determine a slope, we would obtain a slope of –3. Using (0, 87) for the initial reading or vertical intercept (y-intercept), we might find T = –3M + 87 to serve as the predictor equation for this particular set of data. This equation might then be graphed on the graphing calculator over the points already plotted to test for possible fit. If the line is too far from most points, students might select new points for finding another slope or another vertical intercept.

(Continued)

Discuss with students whether a linear graph is appropriate for a cooling mug of tea. Is a period of 6 minutes long enough to adequately observe the cooling tea? Will the tea cool all the way down to 0 degrees Celsius if the mug sits on the counter long enough? Discuss the importance of room temperature in such situations. The tea can cool down only as low as its surrounding temperature, no matter how long the mug sits on the countertop. This is a good example of "misleading data." In addition, if a large insulated pitcher of hot tea were used instead, the insulation would slow the cooling process. As a result, the slope would still be negative but it would not be so steep as the slope of the original data.

NAME _____   DATE _____

10.6

Work with a partner. You will need a graphing calculator.

Situation: A ceramic mug containing tea was heated in a microwave oven, then placed on the kitchen counter and allowed to cool. Its temperature was measured every minute for six minutes and the following data were obtained:

Minutes Out of Oven	Temperature (C°)
0	87
1	83
2	80
3	78
4	75
5	72
6	70

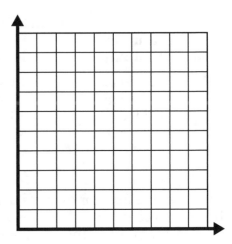

1. Label the axes, select appropriate scales, and plot the data pairs on the grid provided; also plot the data pairs on your graphing calculator grid (select an appropriate window size for the data).

2. Look at the scatter plot and decide on a graphical model that best fits the data. What type of model would you choose?

3. Find an equation that describes your graph best. Be able to give a reason for your specific choice of equation. Graph your equation on the graphing calculator to see how well it fits the data points. Make adjustments in the equation if necessary. (Equations and reasons will vary.) What is your equation?

10.7

Work with a partner. You will need a graphing calculator. The function used in this activity may be a new one for you to explore.

Situation: In an actual science fair experiment, different-sized cans were floated in a bucket of water and the water level was marked on each can. Then a 110-gram weight was placed in each can and the additional amount the cans sank in the water was measured. The following data were obtained:

Can	Diameter of Can Base	Depth Can Sank After Weight Was Added
A	15.6 cm	1.0 cm
B	10.8 cm	1.3 cm
C	7.9 cm	2.2 cm
D	6.8 cm	3.5 cm
E	5.4 cm	6.2 cm

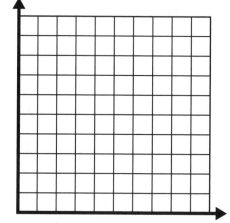

1. Label the axes for diameter and depth change, select appropriate scales, and plot the data as points on the provided grid. The data represent an *inverse variation* of the form $y = k/x$, where $x \neq 0$. You must find an appropriate value for k. This can be done in a variety of ways. For example, use each ordered pair (x_i, y_i) to obtain $k_i = x_i y_i$ for $i = 1, 2, 3, 4, 5$. Use one of these k_i values in the final equation or average all of them together for a final k choice. What is your initial k choice?

2. Once you have an equation to try, graph it on the graphing calculator (select an appropriate window size for the data). Also plot the ordered pairs from the table on the same calculator grid to compare their "fit" to your new graph. Adjustments in k may be necessary. Changes in k cause "bending" changes in the curve. To *raise* or *lower* the curve, a constant b will need to be added: $y = k/x + b$. To move the curve *left* or *right* on the grid, a constant c will need to be added to the x itself: $y = k/(x + c) + b$. Be ready to justify your equation choice to the entire class. What is your final equation choice?

10.8

Work with a partner. You will need graphing calculators and colored markers. For each exercise, choose an appropriate window size for the graphing calculator and appropriate scales for the grid provided.

As you graph each function from the same exercise, make a quick sketch of each curve on the grid provided, using a different color for each curve. Label each curve a, b, c, and so on, as given in the exercise. Record any general graphical changes you observe in each exercise. Be ready to compare the graphical changes in each exercise to the graph of the "parent" function if it is involved.

1 Graph the following set of functions:

a. $y = -3x$
b. $y = -3x + 2$
c. $y = -3x + 8$

Observations:

_____

_____

_____

_____

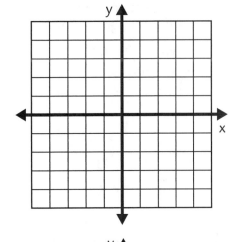

2 Graph the following set of functions:

a. $y = x$ "parent form"
b. $y = x - 2$
c. $y = x - 7$

Observations:

_____

_____

_____

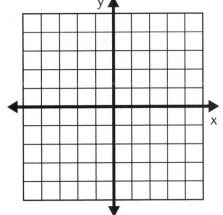

10.9

Work with a partner. You will need graphing calculators and colored markers. For each exercise, choose an appropriate window size for the graphing calculator and appropriate scales for the grid provided.

As you graph each function from the same exercise, make a quick sketch of each curve on the grid provided, using a different color for each curve. Label each curve a, b, c, and so on, as given in the exercise. Record any general graphical changes you observe in each exercise. Be ready to compare the graphical changes in each exercise to the graph of the "parent" function if it is involved.

❶ Graph the following set of functions; notice changes in the coefficients of x:

a. $y = x$ "parent form"
b. $y = 4x$
c. $y = 10x$

Observations:

_____

_____

_____

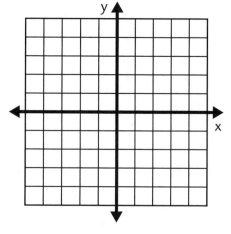

❷ Graph the following set of functions; notice changes in the coefficients of x:

a. $y = x$ "parent form"
b. $y = (3/4)x$
c. $y = (1/5)x$

Observations:

_____

_____

_____

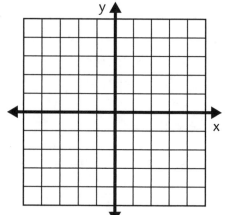

NAME _____ DATE _____

10.10

Work with a partner. You will need graphing calculators and colored markers. For each exercise, choose an appropriate window size for the graphing calculator and appropriate scales for the grid provided. In this activity, you will be exploring the absolute value function: $y = |Ax + B| + C$, where A, B, and C are real numbers.

As you graph each function from the same exercise, make a quick sketch of each curve on the grid provided, using a different color for each curve. Label each curve a, b, c, and so on, as given in the exercise. Record any general graphical changes you observe in each exercise. Be ready to compare the graphical changes in each exercise to the graph of the "parent" function if it is involved.

❶ Let A = 1, B = 0; vary C.

   a. $y = |x| - 3$
   b. $y = |x| + 0$ "parent form"
   c. $y = |x| + 2$

Observations:

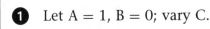

_____

_____

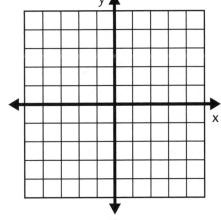

❷ Let A = 1, C = 0; vary B.

   a. $y = |x + (-2)|$
   b. $y = |x + 0|$ "parent form"
   c. $y = |x + 3|$

Observations:

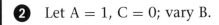

_____

_____

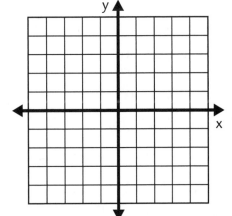

10.11

Work with a partner. You will need graphing calculators and colored markers. For each exercise, choose an appropriate window size for the graphing calculator and appropriate scales for the grid provided. In this activity, you will be exploring the quadratic function: $y = Ax^2 + Bx + C$, where A, B, and C are real numbers.

As you graph each function from the same exercise, make a quick sketch of each curve on the grid provided, using a different color for each curve. Label each curve a, b, c, and so on, as given in the exercise. Record any general graphical changes you observe in each exercise. Be ready to compare the graphical changes in each exercise to the graph of the "parent" function if it is involved.

❶ Hold A and B constant; vary C.

    a. $y = 2x^2 + 5x - 3$
    b. $y = 2x^2 + 5x + 0$
    c. $y = 2x^2 + 5x + 5$

Observations:

_____

_____

_____

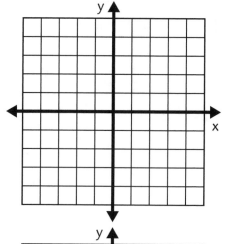

❷ Vary only A for A < 0; B = C = 0.

    a. $y = -3x^2$
    b. $y = -x^2$
    c. $y = -0.2x^2$

Observations:

_____

_____

_____

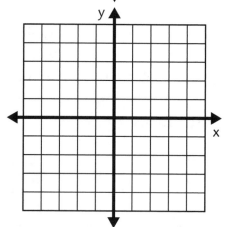

NAME _____ DATE _____

10.12

Work with a partner. You will need graphing calculators and colored markers. For each exercise, choose an appropriate window size for the graphing calculator and appropriate scales for the grid provided. In this activity, you will be exploring the quadratic function: $y = Ax^2 + Bx + C$, where A, B, and C are real numbers.

As you graph each function from the same exercise, make a quick sketch of each curve on the grid provided, using a different color for each curve. Label each curve a, b, c, and so on, as given in the exercise. Record any general graphical changes you observe in each exercise. Be ready to compare the graphical changes in each exercise to the graph of the "parent" function if it is involved.

❶ Vary only A for A > 0; B = C = 0.

    a. $y = 2x^2$

    b. $y = x^2$ "parent form"

    c. $y = 0.2x^2$

Observations:

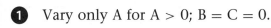

_____

_____

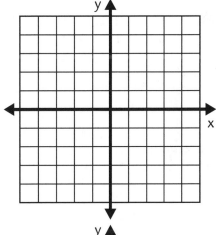

❷ Vary only B for B ≤ 0.

    a. $y = x^2 - 7x + 3$

    b. $y = x^2 - 4x + 3$

    c. $y = x^2 + 0x + 3$

Observations:

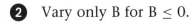

_____

_____

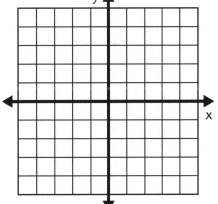

10.13

Work with a partner. You will need graphing calculators and colored markers. For each exercise, choose an appropriate window size for the graphing calculator and appropriate scales for the grid provided. In this activity, you will be exploring general polynomial functions of the type $y = x^n$, for some integer $n \geq 0$.

As you graph each function from the same exercise, make a quick sketch of each curve on the grid provided, using a different color for each curve. Label each curve a, b, c, and so on, as given in the exercise. Record any general graphical changes you observe in each exercise. Be ready to compare the graphical changes in each exercise to the graph of the "parent" function if it is involved.

① Compare these "parent" functions of different powers:

a. $y = x^2$
b. $y = x^3$
c. $y = x^4$
d. $y = x^5$

Observations:
(Hint: How many turns are in each curve?)

_____

_____

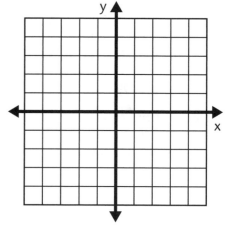

② Compare the following set of functions to their own "parent" functions:

a. $y = x^3 + 5$
b. $y = x^4 - 3$
c. $y = x^5 + 4$

Observations:
(Hint: How many turns are in each curve?)

_____

_____

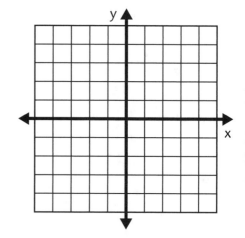

NAME _____ **DATE** _____

10.14

Work with a partner. You will need graphing calculators and colored markers. For each exercise, choose an appropriate window size for the graphing calculator and appropriate scales for the grid provided. In this activity, you will be exploring general polynomial functions of the type $y = x^n$, for some integer $n \geq 0$.

As you graph each function from the same exercise, make a quick sketch of each curve on the grid provided, using a different color for each curve. Label each curve a, b, c, and so on, as given in the exercise. Record any general graphical changes you observe in each exercise. Be ready to compare the graphical changes in each exercise to the graph of the "parent" function if it is involved.

❶ Compare the graphs as the coefficients change.

   a. $y = 5x^3$
   b. $y = x^3$ "parent form"
   c. $y = 0.2x^3$
   d. $y = -x^3$

Observations:

_____

_____

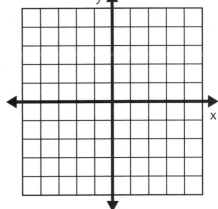

❷ Graph the following set of functions:

   a. $y = x^3 + 2$
   b. $y = x^3$ "parent form"
   c. $y = x^3 - 2$

Observations:

_____

_____

_____

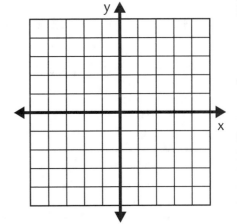

10.15

Work with a partner. You will need graphing calculators and colored markers. For each exercise, choose an appropriate window size for the graphing calculator and appropriate scales for the grid provided. In this activity, you will be exploring general polynomial functions of the type $y = x^n$, for some integer $n \geq 0$.

As you graph each function from the same exercise, make a quick sketch of each curve on the grid provided, using a different color for each curve. Label each curve a, b, c, and so on, as given in the exercise. Record any general graphical changes you observe in each exercise. Be ready to compare the graphical changes in each exercise to the graph of the "parent" function if it is involved.

❶ Graph the following set of functions:

    a. $y = x^3 + 4x$
    b. $y = x^3 + 0x$ "parent form"
    c. $y = x^3 - 4x$

Observations:

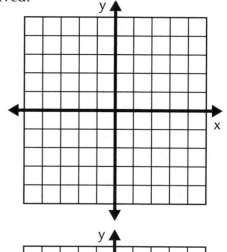

_____

_____

_____

❷ Graph the following set of functions:

    a. $y = x^3 + 3x^2$
    b. $y = x^3 + 0x^2$ "parent form"
    c. $y = x^3 - 3x^2$

Observations:

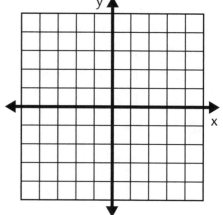

_____

_____

_____

Copyright © 2010 by John Wiley & Sons, Inc. *The Algebra Teacher's Activity-a-Day*

SUGGESTED RESOURCES

BOOKS

Duerr, Elyce. *Algebra Models—Teacher's Guide*. Bloomington, IL: Classroom Products, 2005.

Erickson, Tim, and Rose Craig. *Get It Together: Math Problems for Groups, Grades 4–12*, 11th ed. Berkeley: University of California, 2005.

Lawrence, Ann, and Charlie Hennessey. *Lessons for Algebraic Thinking Series, Grades 6–8*. Sausalito, CA: Math Solutions Publications, 2002.

Martin, Hope. *Daily Warm-Ups: Algebra*. Portland, ME: Walch, 2003.

Martin, Hope. *Daily Warm-Ups: Algebra II*. Portland, ME: Walch, 2004.

Mower, Pat. *Algebra Out Loud, Grades 6–12*. San Francisco: Jossey-Bass, 2003.

Muschla, Gary Robert, and Judith Muschla. *Algebra Teacher's Activities Kit, Grades 6–12*. San Francisco: Jossey-Bass, 2003.

Pilone, Tracey, and Dan Pilone. *Head First Algebra: A Learner's Guide to Algebra I*. Sebastopol, CA: O'Reilly Media, 2009.

Purdy, Scott. *Hands On Algebra, Kindergarten Through Grade Nine*. Solvang, CA: Hands On, 2006.

Thompson, Frances M. *Five-Minute Challenges for Secondary School*. Hayward, CA: Activity Resources, 1988.

Thompson, Frances M. *Hands-On Algebra! Ready-to-Use Games and Activities for Grades 7–12*. San Francisco: Jossey-Bass, 1998.

Thompson, Frances M. *Math Essentials: High School Level*. San Francisco: Jossey-Bass, 2005.

Thompson, Frances M. *Math Essentials: Middle School Level*. San Francisco: Jossey-Bass, 2005.

Thompson, Frances M. (Ed.). *Module 24: Algebra, Grades 9–12*. Austin: Texas Education Agency, 1989.

Thompson, Frances M. *More Five-Minute Challenges for Secondary School*. Hayward, CA: Activity Resources, 1992.

WEB SITES

http://nlvm.usu.edu/en/nav/vlibrary.html

http://archives.math.utk.edu/k12.html

http://www.mathpower.com

http://www.tsbvi.edu/math

http://www.visi.com

http://www.exploremath.com

http://www.lpb.org/education/classroom/itv/algebra/lessons.html

http://illuminations.nctm.org

http://www.borenson.com

http://www.algebrahelp.com

http://www.algebra.com

http://www.pbs.org/teachersource/math.htm

http://mathforum.org

http://www.purplemath.com

http://www.algebracomplete.com

http://standards.nctm.org/document

http://www.webmath.com

http://www.shodor.org/interactivate/lessons/index.html#num

ANSWER KEY

SECTION 1: WHAT DOESN'T BELONG?

Letters of suggested items are shown with reasons why each suggested item differs from the other three items given in the problem; other items or reasons may be possible.

1.1	(c) no m; (d) not metric term
1.2	(c) only one variable; (d) x not squared or coefficient is +1
1.3	(b) has square root sign, or numbers not integers; (c) all even numbers, or does not satisfy Pythagorean theorem
1.4	(c) both sides simplify to b/c; (d) uses binomials or uses addition/subtraction or cannot be simplified
1.5	(a) no constant; (c) coefficients not +1; (d) x is squared
1.6	(a) coefficient not multiple of 2, or has square root sign, or not of even degree; (c) has two variables x and y
1.7	(b) not an inverse variation, or xy product not possible; (c) shows constant, +1; (d) has two constants, c and k
1.8	(b) only one variable, or denominator is +1 when simplified with only positive exponents; (c) simplified form shown on left; (d) rational group cubed, not squared
1.9	(b) only positive exponent shown; (c) simplifies to 14th power in denominator
1.10	(a) no factors outside radical; (c) radicand completely simplified; (d) does not equal other three expressions when simplified
1.11	(a) no x variable, or cannot be factored; (c) only one variable

1.12	(b) not perfect square; (d) quadratic coefficient not +1
1.13	(a) coefficient not prime; (d) no x variable, or when only positive exponents are used, denominator contains no variables
1.14	(a) not cubed form; (d) no variable, number only
1.15	(b) not equivalent to other three; (c) need distributive property to simplify group in brackets; (d) x not first term in binomial
1.16	(a) linear term's coefficient is positive; (c) not perfect square; (d) quadratic term's coefficient not +1
1.17	(c) not factorable with real constants, or x has imaginary value; (d) quadratic term's coefficient not +1
1.18	(b) constant is unit, or uses variable y instead of x; (c) not perfect square, or has negative constant
1.19	(a) has squared terms in denominator; (b) numerator and denominator use only addition; (c) no factor (a + b) remains in final quotient, or cubes used in numerator
1.20	(a) one number not perfect square; (b) not in standard form for conic; (c) is ellipse, not hyperbola

SECTION 2: WHAT'S MISSING?

Arrow orientation shows direction of change made; one arrow points from one item to its partner to represent a change that has occurred or some characteristic that is emphasized; the second arrow with the question mark must apply the same change or emphasize a similar characteristic between its two connected items; suggested expressions for the question marks are provided; other expressions may be possible, depending on which changes are identified.

2.1	$-9(2x)$
2.2	$3y^2 - 15$
2.3	$+3$
2.4	$+0.91$

2.5	$(2c)/b + (4b)/c$
2.6	$1/(3x)$
2.7	$+4y^2$
2.8	$+16$
2.9	$(5y^3)/(x^2)$
2.10	$3/(7x)$
2.11	$+8$
2.12	-10
2.13	$(3x - 1)^2$
2.14	$\sqrt{(x^8)}$ or x^4
2.15	$m^3 - 27$
2.16	$(x^2 + 4)(x^2 - 4)$
2.17	$4(y + w)^2$
2.18	$(x + 3)^2$ or $x^2 + 6x + 9$
2.19	$+3, +1/2$
2.20	$4a^2 + 12ab + 9b^2 + 16ac + 24bc + 20ad + 30bd + 16c^2 + 40cd + 25d^2$

SECTION 3: WHERE IS IT?

Correct item's box number given; selected item must satisfy all clues given in problem.

3.1	7
3.2	3
3.3	8
3.4	6

3.5	5
3.6	4
3.7	1
3.8	3
3.9	2
3.10	4
3.11	9
3.12	1
3.13	8
3.14	3
3.15	8

SECTION 4: ALGEBRAIC PATHWAYS

Correct answer given with possible paths indicated by box numbers; other paths may be possible.

4.1	5.3 feet; path 2-4-7-answer or path 3-5-7-answer
4.2	5 units; path 1-4-8-answer or path 1-6-9-answer
4.3	1/729; path 1-5-8-answer or path 2-6-8-answer
4.4	y^{13}; path 1-5-7-answer or path 3-5-7-answer
4.5	6bc; path 2-1-5-6-10-9-answer or path 4-7-11-12-answer
4.6	1 + 3x; path 1-3-6-8-answer or path 5-9-answer
4.7	+8; path 1-4-8-answer or path 3-5-7-answer
4.8	+11; path 1-5-9-answer or path 2-4-8-answer
4.9	+5; path 1-4-8-answer or path 2-6-9-answer or path 3-5-7-answer or path 3-6-9-answer

4.10	3 < x or x > 3 (graph of inequality on number line not shown); path 1-2-4-8-answer or path 3-6-5-7-answer or path 3-6-5-9-answer
4.11	−5 < x or x > −5 (graph of inequality on number line not shown); path 2-3-7-8-answer or path 1-4-5-6-9-answer
4.12	+7; path 1-4-8-answer or path 2-6-8-answer or path 2-6-9-answer
4.13	+200; path 1-6-9-14-answer or path 2-7-12-15-answer
4.14	−4 or +2; path 2-5-9-14-15-answer or path 2-5-6-11-15-answer or path 4-7-10-13-answer or path 4-8-12-15-answer
4.15	0 or −6; path 1-4-5-7-answer or path 3-8-answer
4.16	+4/3 or −4/3; path 1-4-8-15-answer or path 1-12-15-answer or path 1-11-13-14-15-answer or path 2-10-12-15-answer or path 2-6-14-15-answer
4.17	−3 or −9; path 4-7-9-16-answer or path 2-1-6-10-15-14-answer or path 2-5-8-12-11-10-15-14-answer or path 2-5-8-12-13-15-14-answer
4.18	+3; path 1-6-10-answer or path 3-6-10-answer or path 3-8-12-answer
4.19	+5 or −7, so need +5 and +7; path 3-5-9-8-answer or path 3-5-4-7-8-answer
4.20	x = −1 and y = +1, so (x, y) = (−1, +1); path 1-5-6-13-14-16-answer or path 4-8-12-11-16-15-13-answer

SECTION 5: SQUIGGLES

Suggested solutions shown are in the form of expressions assigned to numbered points; other solutions are possible.

5.1	(1) 5, (2) 8, (3) 9, (4) −2, (5) −4, (6) −25, (7) 27, (8) 2
5.2	(1) −3/4, (2) $\sqrt{2}$, (3) −9, (4) 1.783129..., (5) π, (6) $\sqrt{36}$
5.3	(1) 17, (2) 4, (3) 25, (4) 10, (5) 5, (6) 21, (7) 9
5.4	(1) $\sqrt{61}$, (2) $\sqrt{72}$, (3) $\sqrt{51}$, (4) $\sqrt{99}$, (5) $\sqrt{27}$, (6) $\sqrt{30}$, (7) $\sqrt{25}$, (8) $\sqrt{13}$, (9) $\sqrt{83}$

5.5	(1) 2m, (2) 2mk, (3) k, (4) 8k, (5) 5mk, (6) mk, (7) $6mk^2$
5.6	(1) 12m, (2) 9m, (3) 3m, (4) −6m, (5) −3m, (6) 6m, (7) −9m
5.7	(1) $5x^2y$, (2) ab^2c, (3) mb/2, (4) $8x^3w^2$, (5) 12, (6) −3x, (7) $3w^2$
5.8	(1) $2\sqrt{x}$, (2) 4x, (3) 1, (4) $64x^3$, (5) $16x^2$, (6) $(4x)^{1/3}$
5.9	(1) πr^2, (2) $(1/3)\pi r^2h$, (3) lwh, (4) bh/2, (5) $(1/2)h(B + b)$, (6) bh, (7) πr^2h, (8) $2\pi r$
5.10	(1) $3x − y = 1$, (2) $y + 2x − 3 = 0$, (3) $y = 3x + 4$, (4) $4y = 3x + 8$, (5) $−3x + 4y − 20 = 0$, (6) $2y + x = −2$
5.11	(1) $x + 3y = 21$, (2) $y = x + 7$, (3) $4x − y = −7$, (4) $8x − 2y = 1$, (5) $y = 4x − 5$, (6) $2x + 6y = −30$, (7) $6y = −2x − 3$, (8) $y = x − 5$
5.12	(1) $\sqrt{9m^2}$, (2) $\sqrt{5}y$, (3) $^3\sqrt{8x^6}$, (4) $(27y^9)^{1/3}$, (5) $\sqrt{13}xy$, (6) $^3\sqrt{10x^2}$, (7) $^4\sqrt{16b^8}$, (8) $\sqrt{6abc}$
5.13	(1) 6y, (2) $3x^2$, (3) 25, (4) 5x, (5) 18, (6) 2xy, (7) 12
5.14	(1) 5y, (2) 2x − 2, (3) $5y^2$, (4) $3x^2 + 6x$, (5) $x^2 + x − 6$, (6) 2xy − 2y, (7) 2x + 6, (8) 3x, (9) $x^2 + 2x − 3$
5.15	(1) 2, (2) 2x + 10, (3) $2x^2 + 6x − 20$, (4) 2x − 4, (5) $x^2 + 3x − 10$, (6) x + 5
5.16	(1) $4x^2 − 16$, (2) $x^2 − x − 6$, (3) $x^2 + 2x$, (4) $9x^2$, (5) $3x^2 + 6x$, (6) $5x^2 − 15x$, (7) 4x − 12
5.17	(1) $2d^2 − 6d − 20$, (2) $d^2 + 3d + 2$, (3) $d^2 − 2d − 3$, (4) $d^2 − d − 6$, (5) $d^2 − 5d + 6$, (6) $2d^2 + 8d + 8$
5.18	(1) $4x^2 − 16$, (2) $x^2 + x − 6$, (3) $x^2 + 2x − 3$, (4) 5x − 5, (5) $x^2 + 3x$, (6) x + 3, (7) $x^2 − 2x − 15$
5.19	(1) $y = x^2 + 4$, (2) $y = (−1/4)x^2 + 4$, (3) $y = (1/2)x^2 − 2$, (4) $y = x^2 − 4$, (5) $y = −x^2 + 4$
5.20	(1) $(x + 2)^2 + (y − 3)^2 = 1$, (2) $x^2 + y^2 − 6y + 4 = 0$, (3) $(x − 2)^2 + (y − 3)^2 = 2$, (4) $(x − 2)^2 + y^2 = 9$, (5) $x^2 + y^2 − 4x + 4y + 7 = 0$, (6) $x^2 + y^2 + 4y + 1 = 0$, (7) $x^2 + y^2 = 4$, (8) $(x + 3)^2 + y^2 = 4$

Answer Key